Latino City

American cities are increasingly turning to revitalization strategies that embrace the ideas of new urbanism and the so-called creative class in an attempt to boost economic growth and bring prosperity to downtown areas. These efforts stir controversy over residential and commercial gentrification of working class, ethnic areas.

Spanning forty years, *Latino City* provides an in-depth case study of the new urbanism, creative class, and transit-oriented models of planning and their implementation in Santa Ana, California, one of the United States' most Mexican communities. It provides an intimate analysis of how revitalization plans re-imagine and alienate a place, and how community-based participation approaches address the needs and aspirations of lower-income Latina/o urban areas undergoing revitalization. The book provides a critical introduction to the main theoretical debates and key thinkers related to the new urbanism, creative class, and transit-oriented models of urban revitalization. It is the first book to examine contemporary models of choice for revitalization of US cities from the point of view of a Latina/o-majority central city, and thus initiates new lines of analysis and critique of models for Latina/o inner city neighborhood and downtown revitalization in the current period of socio-economic and cultural change.

Latino City will appeal to students and scholars in urban planning, urban studies, urban history, urban policy, neighborhood and community development, central city development, urban politics, urban sociology, geography, and ethnic/Latina/o Studies, as well as practitioners, community organizations, and grass-roots leaders immersed in these fields.

Erualdo R. González is Associate Professor in the Department of Chicana and Chicano Studies at California State University, Fullerton, USA. His research and teaching interests are community development and participation, urban politics and governance, urban planning and health equity, and Latino urbanism. He examines the intersection of these topics with race, ethnicity, class, and immigration, with an emphasis on Mexican, Chicana/o, and Latina/o communities.

Routledge Studies in Urbanism and the City

This series offers a forum for original and innovative research that engages with key debates and concepts in the field. Titles within the series range from empirical investigations to theoretical engagements, offering international perspectives and multidisciplinary dialogues across the social sciences and humanities, from urban studies, planning, geography, geohumanities, sociology, politics, the arts, cultural studies, philosophy and literature.

For a full list of titles in this series, please visit www.routledge.com/series/RSUC

1 **Shrinking Cities**
Understanding Urban Decline in the United States
Russell C. Weaver, Sharmistha Bagchi-Sen, Jason C. Knight and Amy E. Frazier

2 **Mega-Urbanization in the Global South**
Fast Cities and New Urban Utopias of the Postcolonial State
Edited by Ayona Datta and Abdul Shaban

3 **Green Belts**
Past; Present; Future?
John Sturzaker and Ian Mell

4 **Spiritualizing the City**
Agency and Resilience of the Urban and Urbanesque Habitat
Edited by Victoria Hegner and Peter Jan Margry

5 **Latino City**
Urban Planning, Politics, and the Grassroots
Erualdo R. González

6 **Rebel Streets and the Informal Economy**
Street Trade and the Law
Edited by Alison Brown

Latino City

Urban Planning, Politics, and the Grassroots

Erualdo R. González

LONDON AND NEW YORK

First published in paperback 2018

First published 2017
by Routledge
2 Park Square, Milton Park, Abingdon, Oxon OX14 4RN

and by Routledge
711 Third Avenue, New York, NY 10017

Routledge is an imprint of the Taylor & Francis Group, an informa business

British Library Cataloguing-in-Publication Data
A catalogue record for this book is available from the British Library

Library of Congress Cataloging-in-Publication Data
Names: Gonzâalez, Erualdo R., author.
Title: Latino city : urban planning, politics, and the grassroots /
Erualdo R. Gonzâalez.
Description: Abingdon, Oxon ; New York, NY : Routledge, 2017. | Includes index.
Identifiers: LCCN 2016038457| ISBN 9781138820548 (hardback) | ISBN 9781315743806 (ebook)
Subjects: LCSH: City planning—Social aspects—California—Santa Ana. |
City planning—Citizen participation—California—Santa Ana. |
Urban renewal—Social aspects—California—Santa Ana. |
Urban renewal—Citizen participation—California—Santa Ana. |
Hispanic Americans—California—Santa Ana—History. | Santa Ana (Calif.)—Social conditions.
Classification: LCC HT168.S238 G66 2017 | DDC 307.1/2160979496—dc23
LC record available at https://lccn.loc.gov/2016038457

ISBN: 978-1-138-82054-8 (hbk)
ISBN: 978-1-138-59549-1 (pbk)
ISBN: 978-1-315-74380-6 (ebk)

Typeset in Times New Roman
by diacriTech, Chennai

Contents

Figures

Preface

"El centro de la ciudad de Santa Ana emula a muchas ciudades de Mexico. La Calle Cuarta, entre la Main y Broadway, es el area más comercial de esta urbe de Orange ... y donde la marca es: Hecho en México." [Downtown Santa Ana emulates many cities in Mexico. Fourth Street, between Main and Broadway, is the largest commercial area of this Orange metropolis ... and where the slogan is: Made in Mexico] (Almada 2005).

– From *La Opinion*, the largest Spanish language newspaper in the United States based out of Santa Ana's neighboring Los Angeles.

"Hay palabras que duelen, especialmente cuando se dirigen contra nuestra raza. Aqui se ha dicho que se piensa traer gente buena y yo me pregunto si los Latinos, especialmente los Mexicanos, '¿Somos gente mala?' *"* ["There are words that hurt, especially when they are directed toward our people. It has been said here that the idea is to bring good people [to the downtown] and I ask myself if Latinos, especially Mexicans, 'Are we bad people?'"

– Daniel Estrada, downtown business owner (Semanario Azteca 1984).

What began as a protest against police brutality in downtown Santa Ana in Southern California on the evening of August 15, 2014 exploded into a heated gentrification confrontation between protesters and barbers from American Barbershop on 4th Street—pitting "Mexicans against Mexicans," said the *OC Weekly* (San Roman 2014). The activists, leaving a rally they held at Sascer Park, were returning to the eastern part of downtown where the group had originally assembled, right across from the barbershop. Their path had taken the activists by a concentration of businesses that have since the 1970s served Mexican working-class immigrants and newer restaurants and hip boutiques geared for the upwardly mobile. Some activists were chanting "Gentrification sucks!" and "Sell-outs" at patrons at one of the newer restaurants when a near-melee happened in front of the barbershop (ibid., p. 1). Some activists, many of whom were Mexican and native born of Mexican descent, or Chicana/o, maintain that some of the barbers had flicked cigarette butts at them and yelled "Wetbacks!" (ibid., p. 1). George Mendoza, owner of the barbershop, denies this account, saying that protestors taunted without provocation, "'Fuck American Barbershop. Fuck you and your little hipster haircuts" (ibid., p. 1). "We are all Mexican here! For God's sake!," he told the

O.C. Weekly, adding that his parents came to the U.S. "illegally." He called the protestors "a bunch of dumbshit, fucking ignorant people," concluding "it makes me embarrassed for my own people" (ibid., p. 1).

Events like this echo and amplify the rhetoric I heard four years earlier in city hall. On the evening of May 20, 2010, members of the Santa Ana Collaborative for Responsible Development (SACReD) assembled in Santa Ana City Hall to negotiate twenty-eight demands for a legally mandating community benefits agreement (CBA), between the city, a developer, and the community for a portion of the downtown area. The meeting was with city officials, of which two were city councilmembers, and a developer. About sixty people attended: over thirty SACReD members and ten of their allies, attorneys for both the city and the collaborative, and Community Development Agency senior staff, among other city staff and community members.

About a half hour into the start of the meeting, Sal Tinajero, a member of the first all-Latina/o City Council of a major city in the United States, abruptly and lamentably dismissed himself. The councilmember said he had not been able to review the collaborative's revisions to the CBA draft and was therefore unprepared to discuss them, the meeting had too many people, and speaking too much could potentially create negative consequences for him or the city. The collaborative had provided him the demands about a week prior. Yet, he also acknowledged that the negotiations were very important because no grassroots community group in the city's history had ever worked on neighborhood planning so proactively and thoroughly to negotiate a proposed legally binding agreement between the city, developer, and the community. He and other councilmembers had acknowledged this in other public settings before.

But, there was also a high level of discomfort in the room tied to a history of racial and class tensions between some SACReD activists representing two neighborhoods. Councilmember Michelle Martinez pointed out, authoritatively and confused, that some SACReD members' personal agendas go against SACReD's demand for more affordable housing units catering to a lower-income population than the city and developers had planned for. "They don't want affordable housing. They don't want stuff on lot 1 and that is majority for [proposed] low income housing and that should tell you that they don't have interest in your purpose," she said. Most in the room understood that she was referring to a historic neighborhood group in a majority White middle-class neighborhood that had relatively recently joined SACReD. In the end, these negotiations were tabled for another and smaller meeting.

Ethnic, class, and political tensions have long encompassed battles over community development in Santa Ana. This book explores how city redevelopment plans and related city documents, as well as broader media and local publications, socially construct problems and imagined communities, erase existing populations from redevelopment thought, and create a set of practices prioritizing future communities, using Santa Ana as a case study. It also addresses how the grassroots and residents go about challenging rather than accepting such processes, the differences that some of these efforts make, and challenges and contradictions that they encounter along the way. That is, this book is also about showing how the grassroots goes about identifying and narrowing what Brenner (2012) refers to as the disjunction between the actual and the possible.

I focus my analysis on Santa Ana in Southern California from the early 1970s to the early 2010s. I take a critical and temporal look at the extent of historical specificity of community development discourse and practice and how it helps us to understand, or not, recent events in Santa Ana. This temporal and critical approach provides a useful lens for examining how discourse and practice may be reinvented in relation to racial and political-economic conditions and urban redevelopment trends, and may maintain core intentions. At the grassroots level, this approach lends to examining the types of conflicts in the broader community provoked from redevelopment discourse and practice and how the broader community responds via community-based approaches to change. I show how activists have waged a number of community-based struggles over proposed and actual community development and taken on Santa Ana city hall and developers.

Through four decades and changing forms, from an open-air mall cultural-themed redevelopment to attracting the creative class and embracing new urbanism, this book traces how the discourse and practice of redevelopment in Santa Ana have routinely spatially alienated Latina/o, the working class, and immigrant residents. Instead of viewing redevelopment plans, controversies, and grassroots struggles in isolation, I analyze them critically as connected and part of a larger narrative over race and ethnicity; immigration, especially from Mexico; class; privilege, community redevelopment, and political economy; and grassroots activism. *Latino City* provides an in-depth and interdisciplinary case study that weaves these concerns together.

How I Came to Write This Book

I grew up in Santa Ana. I began this research in 2007 out of an interest in the city's ongoing redevelopment of the downtown. The subject first garnered my active interest in the mid-1990s when local newspaper stories began endorsing the city's inaugural Artists Village, a cultural-led redevelopment project, which consisted of rehabilitated historic buildings, building work-live loft development, and a few art studios, galleries, and cafes. The Artists Village is situated one block adjacent from the downtown's busiest and largest commercial street, Fourth Street, which locals call *La Calle Cuatro*, or *La Cuatro* for short, a very different commercial and cultural world—jewelry stores, shoe stores, bridal/quienceañera stores, clothing stores, and a few restaurants and other amenities marketed to the city's and region's Mexican, working-class, and immigrant community. *La Calle Cuatro* means Fourth Street in Spanish. These are the stores the activists passed in 2010. I spent a lot of my childhood and youth in *La Cuatro*, visiting or spending whole days with my mother at the shoe store that she managed. The area was a very busy place at the time, and everyone I knew valued it as a site of cultural pride and connection. We all feared that the city was planning to crowd that culture out of *La Cuatro*. Quietly, I wondered, who in city hall was making redevelopment decisions and how was the broader Santa Ana resident community incorporated in these redevelopment decisions?

My interest in the questions about the Artists Village deepened when I obtained a degree in urban planning. The ambitious redevelopment projects that followed stoked out my fears. Some of the stories that ran in the local newspapers and on the Internet about redevelopment plans that began to crop up in city government in 2007 explicitly criticized the current redevelopment vision, the Renaissance Specific Plan (RP), in a way 1990s coverage of the Artists Village never had. State actors, locals accused, were aggressively and enthusiastically drafting redevelopment proposals and carrying out redevelopment activity that extended to purchasing and boarding up homes in the downtown area with a primary intention to gentrify and displace the majority Mexican, working-class, and immigrant residents and small businesses that market to them.

I joined SACReD in 2007, joining its fight against the RP. I was a recently minted Ph.D. in Urban and Regional Planning, had just started my position as an Assistant Professor at nearby California State University, Fullerton, and was still living in my native Santa Ana. I was now asking questions resembling those I had discussed a decade earlier with reference to the Artists Village, but with a more fine-tuned theoretical perspective. I believed I could find the answers with SACReD. But my participation was also serving my interest in being an active resident of Santa Ana. I was eager to bring my ten years of experience with community-based and applied research and evaluation to bear in the city I love. I had worked side-by-side with community organizers and community leaders across California on a host of topics, such as environmental justice, community development, and community building. Certainly, some of these experiences and skills could be useful. How could I not try to figure out how?

At first, I wasn't sure how I might apply my academic and community-based research background in SACReD, but I was committed to merging my activism and scholarly publications. As I made the transition to full-fledged Ph.D. and principle investigator of my own original contribution to the field, I understood that SACReD was not a "community research project" where I would enter and collaborate to perform a research task to be completed by a specific deadline or under a particular research grant mandate.

My involvement with the collaborative helped me understand my pathway to writing this book. In the mid-1990s while 35 miles away at college at Loyola Marymount University (LMU), Los Angeles, I nonetheless monitored the creation of the Artists Village. One morning, visiting home for the weekend, I picked up the *Orange County Register* newspaper from my parents' driveway and learned that several councilmembers, local businesspeople, and community leaders were taking trips out of the state and around California to tour similar artist-related redevelopment projects to learn what ideas they could apply locally. None of the coverage seemed to explain the questions that would naturally precede the development itself. Who came up with this vision and why? Who participated in its development?

The deeper roots of this book precede the Artists Village however. My mother grew up near the historic downtown in the late 1960s. My family and my extended maternal side of the family continued to live in Santa Ana when I was a child. Most of us lived within three miles connected by one major street, McFadden. The full scope of my positive memories of *La Cuatro* exceeds what I can express here. My mother, Margarita Romero González, worked in downtown on and off, from the 1970s to the early 1990s, even before the street came to be popularly called *La Cuatro*. The decision to work in and around downtown was an easy choice. My maternal grandparents, Miguel Romero Jimenez and Sara Castro Romero, had brought my mother and her siblings from Mezquital del Oro, Zacatecas, México when she was 12 in 1966. The move was a return for my grandmother, who had been born blocks from Santa Ana's downtown in 1924, although she was raised in Mexico. As a young teen, my mother worked at the Santa Ana Library located in the Civic Center adjacent to downtown and at a record store on *La Cuatro*. My mother routinely references a late 1960s downtown largely frequented by White consumers and by the early 1970s many vacancies in the area.

When I was a school-aged child in the late 1970s, my father, Lucilo González Ramirez, and I visited my mom frequently at *Calzado Canadá*, informally known as *La Canadá*, a Mexican shoe store she managed on the corner of Fourth Street and Main Street, the heart of downtown. My father had crossed the border several times, once in the trunk of a car, about a decade earlier from Huisquilco, Jalisco, in his late teens. He would carry me on his shoulders, and together we would be a giant navigating the sidewalks, providing me a bird's eye view of storefronts, pedestrians, consumers, and streetscapes. By the 1980s and early 1990s, especially during summer periods, I spent a good deal of time with my mother on *La Cuatro*—at the shoe store where she was an employee and manager, the two clothing stores she owned for a short period, and other businesses of employment. When I was not preoccupied drawing palm trees, pyramids, and beaches behind the register counter or playing my favorite records, I would regularly help with general upkeep, organizing shoes in the massive window display and using excessive Windex to clean windows. I took little notice of the sidewalks filled with people, many coming into the store, and the windows reverberating from the loud music from a nearby record store. I assumed *La Cuatro* would always be busy.

During different hours of the day my mother and I would walk the busy sidewalks and she would interact with business owners in the area. There was Pedrini's Music, a large music store where I enrolled in piano classes, owned by Tom and Rosita Pedrini, and Joyeria Latina, owned by Juan and Cecilia Portillo (El Señor y la Señora Portillo), among others. There were also a few eating establishments that we would frequent. Los Panchos, the largest Mexican restaurant in the area, is where my family and I would eat often when my mom was out of work, and Rancho Mendoza is where my mom and I would especially enjoy visiting during lunch for what we believed were the best *sopes de carnitas* in the area. As a boy,

I had little interest in the countless jewelry stores and bridal shops, but they were part of the fabric of the downtown that I treasured.

Just before my teenage years, my time on *La Cuatro* expanded, but by the age of 17, this began to fade. My older cousin, Benjamín Romero Vázquez, and I would meet up during his break from working at Joyeria Latina, which was across from the shoe store. We would walk around, passing by *discotecas* (record stores) playing loud Spanish pop music, often visiting one of the area's last standing mainstream store—Woolworth's—to buy Whoppers candy and eat delicious BLT sandwiches. By the late 1990s, my mother found work opportunities elsewhere and her era on *La Cuatro* ended. To this day, my mother talks about the fondness that she had for *La Cuatro* and the friendships that she created there. When I asked my mother about some of her best memories about working on *La Cuatro*, she recalled: "Los viernes nos ibamos a cenar a Avilas El Ranchito por la calle primera. Ibamos amigas del trabajo y que trabajaban en tiendas de joyería, novia, y ropa, Fue algo bien bonito. ["The girls from work and I, along with other friends who worked in jewelry stores, bridal shops, and clothing stores would meet up and go out to dinner Friday nights. Those were very nice times"]. These early experiences immersed in the sites, sounds, and culture of *La Cuatro*, at what I eventually understood was the area's height, drive my research.

Even at the time, I developed some insights that some Orange County kids perceived Santa Ana in a negative light. For instance, playing sports for my Catholic K-8 school, Immaculate Heart of Mary, required that our team travel to Catholic Schools in more well-off areas of the County and I learned that most opposing teams thought of us, pejoratively, as the Mexicans from Santa Ana. I then had the opportunity to attend Mater Dei High School, the most distinguished school in the County where many of the kids are from economically well-off families from the region. I learned further that many of my new friends and classmates always commuted quickly after school, fearing danger and an unfamiliar city. While my personal experiences in school were overwhelmingly positive, I never got used to the disdain that others who lived outside the city felt for my hometown.

In 1995, I took a summer job for the city of Santa Ana during college as a counselor as part of the workforce development program. I placed youth from working-class and predominant Mexican immigrant neighborhoods like mine in summer internships with city, state, and federal offices located in the Civic Center. I was excited to have a formal work connection to the city. My mother, knowing of my interest, had spoken with city personnel about me, and Roger Kooi, downtown development manager at the time, who would patrol downtown, tipped me to the job after she told him about me.

As an undergraduate at LMU, I took classes focused on city politics, race/ethnicity, community psychology, and program evaluation. The classes focused on city politics routinely included politicians, community organizers, and scholarly guest speakers who covered Latina/o urban questions pertaining to Los Angeles County. Decentralization, deindustrialization, demographic shifts, political representation, activism, applied research, and social justice, among other topics, were common in those classes and I wondered with each passing guest speaker and class lectures how these topics applied to Santa Ana. Those academic questions remained with me for years. By my senior year, I volunteered as an evaluation research assistant to Dr. Cheryl Grills in the Psychology Department. I bridged coursework with community-based evaluation research and continued to work with non-profits and community organizers throughout California from poor and working-class African-American, Native American, and Latina/o urban and rural neighborhoods after graduation. I was exposed to their work around school reform, race relations, and community health, among other topics, and at the core of their efforts was community organizing, leadership development, and advocating for city or school district resources. This education and research experience sparked my interests to analyze cities through critical race and systems perspectives and to understand and question the extent local politics and grassroots organizing could make cities better.

These interests deepened during my Ph.D. training when I continued to live in Santa Ana. My training in the Department of Urban and Regional Planning at the University of California, Irvine prepared me in the areas of community development, public policy, community participation, evaluation research, and Chicana/o Studies. As a research assistant to professors on various projects, I had the opportunity to work with numerous working-class communities of color throughout California and other states. The common element to my work was developing community-based research methods and researching or evaluating the ways in which resident leaders and groups were carrying out environmental justice campaigns and community health projects and any difference these efforts were having on residents and local public policies. My master's thesis topic also focused on the grassroots, examining fear of deportation as a potential barrier to community participation. This factor, which I had observed in non-systematic ways through other community research experiences, was missing from the general community participation literature and by this stage in my academic and my community-based research and evaluation tenure, I had developed an intuitive sense that this may be the case among some undocumented Mexican immigrants. My city lived experiences informed this suspicion and collecting data in my city was an obvious choice. Santa Ana was and continues to be Orange County's most Mexican immigrant city. About half of the City's more than 330,000 residents are foreign born, the vast majority from Mexico, and about half are unauthorized (U.S. Census 2015). Santa Ana has maintained relevance for anti-immigrant discourse and activities. Indeed, in 2008, a *U.S. News and World Report* article characterized Santa Ana as the "poster child" for national immigration issues (Schulte 2008). My dissertation broadened to the non-profit level and examined the extent to which community-based organizations with different missions, those with an advocacy orientation versus those with a consensus mission, were more likely to include resident participation in planning of their projects and share decision-making power with residents when they carried out general community projects.

I joined SACReD in 2007 one year after completing my Ph.D. I had gained a faculty position at CSU Fullerton that same year and this allowed me to live in Santa Ana and make an easy commute to campus. I began reading academic literature to help me understand the downtown area redevelopment models that the city was using. I could map the city's choice model to alter the area's built environment and demographics to a new trend called new urbanism by the approach to constructing housing opportunities, such as work-live lofts, targeting small families. The RP on which we at SACReD focused our efforts would have been the city's most comprehensive and ambitious redevelopment plan for the downtown area in history. I became preoccupied with understanding the steps that the city and the urban designer firm under contract took in drafting the RP, who was involved and left out of the planning, who influenced the planning, and why. I had no desire to once again have a formal work connection to the city or become a politician, as some suggested, to find answers. I wanted to be *of* the grassroots in Santa Ana. I hoped that there would be opportunities to be politically active via the grassroots and apply my community-based research expertise in my city, in and around the downtown area, a place where I had a history. I also aimed to turn my experiences into the book you are now reading, both as a means to live up to my professional obligations as a scholar and to document a set of issues I believe are critical to the future of the Latina/o population in the United States.

Having halted the RP, SACReD began to negotiate a community benefits agreement for a transit-oriented development (TOD) project within the RP. We created our own community-based participation activities and uncovered resident redevelopment concerns that were not accounted for by the city and that developers had ignored. SACReD allowed me first-hand evidence of how institutional processes worked in the city.

Participating in SACReD gave me direct access and a working relationship with housing and community-development professionals working in the non-profit sector as well with seasoned activists who were at the forefront of community organizing against city downtown visions, plans, construction, and demolishment of housing and commercial options since at least the mid-1970s, time periods which I cover in this book.

I could see that discrimination against Mexican, immigrant, and working-class and poor groups had fueled the discourse and practice of downtown redevelopment since at least the 1960s, and I wondered if these sentiments continued to play a role in the actions of many city actors and developers. Not irrelevantly, the *OC Weekly* deemed Santa Ana the second most racist city in Orange County in 2009. The commentary pointed out the city's historical racism such as the destruction by fire of the city's Chinatown, a public lynching of a Mexican in 1906, and several pioneering city leaders who were Klansmen (Arellano 2009). During the 1990s, Orange County was the birthplace of major state and federal anti-immigration and Spanish-language initiatives, Propositions 187 and 227. Proposition 187 was the state ballot initiative—which passed but ultimately was overruled in federal court—to deny state services to unauthorized immigrants, such as health care and education, and required public officials including teachers and health care personnel to report suspected unauthorized immigrants to immigration enforcement. Proposition 227 eliminated bilingual education in California. The Minuteman organization, a citizen vigilante group also based out of Orange County, began monitoring unauthorized immigration at the U.S.-Mexican border. As Chavez (2008), Vallejo (2012), Arellano (2008) and others have argued, the public discourse surrounding these initiatives and organization helped surge a growing anti-Mexican immigrant sentiment in Orange County and the country where Mexican immigrants were inherently a criminal ethnic group, draining public assistance and unwilling to assimilate.

In summary, these personal experiences and broader political and urban concerns help fuse the ensuing chapters. I hope this book will help advance and stimulate future research and more responsible community development policy-making on the subjects of this book.

References

Almada, Morales Jorge. 2005. "Santa Ana tiene sabor a Mexico." *La Opinion*. www.sre.gob.mx/umi/051212.htm. Accessed 15 October 2006.

Arellano, Gustavo. 2008. *Orange County: a personal history*. 1st Scribner hardcover ed. New York: Scribner.

Arellano, Gustavo. 2009. "Most racist city in Orange County—Is it Fullerton??? Our top five." December 8, 2009. www.ocweekly.com/news/most-racist-city-in-orange-county-is-it-fullerton-our-top-five-6470750.

Brenner, Neil. 2012. "What is critical urban theory?" In *Cities for people, not for profit: critical urban theory and the right to the city*, edited by Neil Brenner, Peter Marcuse, and Margit Mayer, 11–23. Abingdon, Oxon: Routledge.

Chavez, Leo R. 2008. *The Latino threat: constructing immigrants, citizens, and the nation*. Stanford, CA: Stanford University Press.

San Roman, Gabriel. 2014. "Mexicans versus Mexicans in downtown Santa Ana gentrification shout-off–then someone yelled 'wetback.'" *OC Weekly*, August 17, 2014. www.ocweekly.com/news/mexicans-vs-mexicans-in-downtown-santa-ana-gentrification-shout-off-then-someone-yelled-wetback-6448703.

Schulte, Bret. 2008. "Mexican immigrants prove slow to fit in." *US News and World Report*, May 15, 2008. www.usnews.com/articles/news/national/2008/05/15/mexican-immigrants-prove-slow-to-fit-in.html.

Semanario Azteca. 1984. "Molestia en comunidad latina ante accion discriminatoria." *Semanario Azteca*, 1, 9.

U.S. Census. 2015. Population estimates, July 1, 2015 (V2015).

Vallejo, Agius Jody. 2012. *Barrios to burbs: the making of the Mexican American middle class*. Stanford, CA: Stanford University Press.

Acknowledgements

A village of family, community, and institutions knowingly or unknowingly helped shape this book. I have deep appreciation for this village because the nature of this book is significantly tied to this support. I am responsible for any misinterpretations of judgment in these pages.

The academic enterprise: Raul P. Lejano patiently and generously read and edited much of my writing building up to this book. He assisted with the book's framework and co-authored with me previous journal articles that inform chapter three of this book. I continue to gain from his wisdom. Rodolfo D. "Rudy" Torres broadened my critical outlook on cities, sparked my confidence to write this book, and guided the book proposal stage for this project. Carolina Sarmiento, Ana Urzúa, and Susan Luévano gently pushed me to refine my thinking about my role in the community and what this means for social justice and scholarship. We co-authored a piece on our community activism in Santa Ana and chapter four builds and expands on that work. Clara Irazábal-Zurita and Melissa Hidalgo were very generous and provided invaluable feedback on the book's Introduction and first chapter. Denise Sandoval, Horacio Ramirez Roque (rest in power), David Alvarez, and Tomás Summers Sandoval welcomed me to their writing group when I was drafting my first manuscript tied to this book. Some suggestions from those sessions certainly made their way to this book. My colleagues in the Department of Chicana and Chicano Studies and the Asian American Studies Program at California State University, Fullerton always expressed confidence in this endeavor. This assurance went a long way and makes this book all that more meaningful to me.

Community: Unequivocally, this book as a whole also stems from my involvement at El Centro Cultural de México (El Centro) in Santa Ana and the friendships, collaborations, and networks that flourished from it. El Centro provided me the space to see how activists genuinely commit to the broader social, political, economic, and cultural good. It was through El Centro that I became involved in the Santa Ana Collaborative for Responsible Development (SACReD), the focus of chapter four in this book. El Centro also led me to meet long-time Santa Ana and downtown activists, historians, leaders, and groups who are part of this book's general story. Sam Romero and Teresa Saldivar always made time to meet with me to review early and more recent drafts of chapters two and three. They were consistently available. Gustavo Arellano reminded me to spruce up the preface! I hope to have understood and accurately represented their views.

Institutional support: I benefited tremendously from a number of research grants that provided support for this study. California State University, Fullerton's (CSUF) Office of Academic Affairs allowed me time and generous support to collect the data that informs this project. Several research awards from CSUF's Office of the President also provided support for writing and editing. I gained tremendously from the Department of Chicana and Chicano Studies that provided conference funding to present and receive feedback on parts of the book. The Department was also the gateway to work with undergraduate students, most former students of mine, who went on to become part of my research team. Anna Diaz Villela, Walter Santizo, Angelica Ruiz, and Alexa Harris collected field data. Georgina Martinez, Christian Orozco, and Joel Mondragon collected archival research. Janet Bernabe and Estevaliz Castro produced an annotated bibliography. I am very grateful to this team. Kelly Donovan, Cara Gafnar, and Lorena Guadiana created the infographics and maps for this book. I also acknowledge Faye Leerink and Priscilla Corbett at Routledge who generously guided me through the publication process. I am very thankful to the Santa Ana Public Library, especially Manny Escamilla in the History Room who directed me to invaluable city redevelopment historical documents and always fielded my queries with enthusiasm.

Family and friends: Most especially, I send *un abrazo* to my family and friends. I am extremely appreciative for my parents, Margarita Romero González and Lucilo González Ramirez, who encourage and support me. My mom *always* made time from her busy work schedule to accommodate this book. Twice a week for two years she drove 45 miles to and from my home. My mom helped care for mi bebé Zenaida Maité who was just an infant when I began to write this book. She also fed me, fueling me weekly with her home cooking. "*Como vas con tu libro*? ("How is your book coming along?"), she always asked. While my answer varied weekly, I knew I could always count on her. *Gracias* mom. I thank my brother, René Romero González, who often checked in on my progress. I partnered with my cousin Benjamín Romero Vázquez via El Centro. Our shared experiences and conversations about our city kept me on my toes to complete this project. Too many to list here, I am very fortunate to have the very strong support of uncles, aunts, cousins, and friends nearby who kindly expressed interest and waited patiently for this book to come to fruition.

I dedicate this book to my *esposa* Connie and *hija* Zenaida Maité. Connie was extremely understanding of the time that I required for this project. Too often, I would scoot out of the house to write very early in the mornings or late in the evening. This made it complicated to maintain routine family and social activities. Connie shouldered responsibilities that we both should have shared during our daughter's first two years of life. My feelings about her support are light years beyond "indebtedness." Zenaida was three months old when I began writing this book. She nourished my soul daily and made writing much easier when I needed a lift. Zenaida, while the content of this book will appear very outdated when you read it (hopefully), many of the book's themes will remain in your lifetime. I dedicate this to you because a message in this book is that we have a role and responsibility ensuring that our world, cities, and neighborhoods be ripe for the most deserving.

Acronyms

CBA	Community Benefits Agreement
CC	Creative Class
CNU	Congress for the New Urbanism
DSADP	Downtown Santa Ana Development Plan
PBID	Property-Based Improvement District
PAR	Participatory Action Research
RP	Renaissance Specific Plan
SACReD	Santa Ana Collaborative for Responsible Development
SANO	Santa Ana Neighborhood Organization
SD	Station District
NU	New Urbanism
TOD	Transit-Oriented Development

Introduction

Urban planning in the Latino city

This book builds on a still modest but growing collection of work focused on Latinas/os and urban and city concerns in American cities, generally referred to as *Latino urbanism*. Since 2000, Latino urbanism has emerged as a theoretical and practical approach to community development and planning policy that seeks to understand how socio-economic and political change affects the nature and functioning of cities with large Latina/o populations over time (Diaz and Torres 2012). Major themes in this body of literature relate to urban politics, political economy, and immigrant communities and local activism over housing and public spaces (Beard and Sarmiento 2014, Irazábal and Farhat 2008, Main and Sandoval 2015, Sandoval 2009). Specifically, this body of work is concerned with understanding the extent that Latina/o urban living, work, shopping, and play preferences and experiences are tied or not to fair and just baselines of opportunities and conditions and the role classist and racialized public policy and civic practices have in these scenarios. While the planning literature has historically ignored Latino urbanism (González and Irazábal 2015), Latina/o urbanists such as Valle and Torres (2000), Davis (2000), and Diaz (2005) have produced a collection of books and addressed urban areas predominated by Chicana/os and Mexican immigrants in the Southwest. These and other scholars helped create the Latino urbanism movement and have made modest progress critiquing conceptual and practical developments within the field of urban planning (Rios, Vazquez, and Miranda 2012, González and Irazábal 2015, Irazábal 2012). Latina/o urbanists are incresingly addressing new urbanism (NU), urban design, creative cities or creative class (CC), and how these models contribute to gentrification (González and Lejano 2009, Diaz 2012, Rios, Vazquez and Miranda, 2012, González and Guadiana 2013, Lara 2012, Londoño 2012, Diaz and Torres 2012).

Contemporary redevelopment trends and key emerging racial-ethnic issues in urban planning are vital contexts for this book (González and Irazábal 2015). Increasingly, in U.S. heavily populated cities at the center of large metropolitan areas, developers, investors, and businesses are collaborating with municipal officials to redevelop and gentrify (Glaeser and Shapiro 2001, Frey 2012). In 2011, for example, the major central cities of the country's largest metropolitan areas grew faster than their combined suburbs (Frey 2012). In fact, one-third of the U.S. population was living in a city, the highest proportion since 1950 (U.S. Census

Bureau 2012a). Between 2000 and 2010, the area within a two-mile radius of the city hall in all the nation's cities with metropolitan areas of at least five million people had experienced double-digit population growth rates (U.S. Census Bureau 2012a).

The changing racial demographics of the urbanization of the United States represents to some extent a reverse of White flight (U.S. Census Bureau 2012a, 2012b). But, the Latina/o population also increased in each of the 20 largest central cities during the same period (U.S. Census Bureau 2012a). These demographic changes are largely attributed to people flocking to the downtown areas of these cities, which are increasingly becoming preferred places to work, live, and play. We can tie this budding attraction to the central city to developers, investors, and businesses who are increasingly collaborating with municipal officials to redevelop and gentrify heavily populated cities at the center of large metropolitan areas (Glaeser and Shapiro 2001, Frey 2012). For instance, between 2000 and 2010, metro areas with five million or more people had experienced double-digit population growth rates within two miles of their largest city's city hall, more than double the rate of these areas overall (U.S. Census Bureau 2012a). At the same time, many central cities with the largest African-American populations, such as Detroit and Chicago, saw a decline in the African-American population. Planners—I use the term broadly, here, to include local elected and appointed officials, city planners, city staff, developers, investors, civic leaders, and urban planning and design consultants—use gentrification to imply middle- and upper-middle-class Whites. These trends are certainly appealing for some central city stakeholders who are mostly interested in "bringing back" people to downtown areas who could afford and reap the benefits of redevelopment.

Because many Latina/o-majority cities are large- and medium-sized cities, it is likely that proposals for redevelopment will encounter existing downtown Latina/o neighborhood and commerce targeted at Latinas/os, just as they have in Santa Ana. Thus, the conflicts with maintaining or setting a fair and just baseline of consumption and living conditions for majority residents in these central city areas—typically the poor, working class, and communities of color—that have been growing for decades in Santa Ana will likely crop up in other cities. This book advances the body of work within the Latino urbanism literature focused on addressing critical questions about present-day redevelopment models for central city downtown areas such as new urbanism (NU), creative city (CC), and transit-oriented development (TOD) in areas with a substantial Latina/o population.

Redevelopment Toward an Ideal Community

While TOD, NU, and CC redevelopment models each have unique goals toward quality of life issues, all have a common implicit or explicit aim of attracting "vibrant" and "diverse" communities. Quite commonly we find that city planners who adopt or wish to apply any one of these models are doing so in areas government and the private sector have largely ignored for the past half a century.

In recent decades, there has been competition among different cities for private investment to carry out these models, largely where municipalities are cash-poor (Harvey 1989, Brenner and Theodore 2002, Grodach and Loukaitou-Sideris 2007).

The central principles of the TOD, NU, and CC models have similar understandings of idealized populations that should primarily benefit from redevelopment. These tenets also underpin most of the thinking of planners in creating seminal projects remaking American cities. Lejano and González (2016) argue that the term *genitocracy* more aptly describes the consumption populations present-day urban planners seek to attract rather than gentry. Rather than a well-defined middle-class, a *genitocracy* is a fluid intersection of race, class, citizenship, culture, and other categories to bear on the idea of privilege (ibid.). In practical terms, planners invoke the *genitocracy* variously through terms such as creative, young people, professionals, and hipsters.

Planners and policy-makers frequently credit TOD, NU, and CC redevelopment as a means to attract idealized groups back to the central city, particularly to downtown areas, to work, live, and consume. As I will explain below, the key ideas that guide the projects bent on capitalizing on a reversal of White flight—the historic pattern of White residents moving to the suburbs—have both commonalities and differences.

Planners typically describe TOD redevelopment as a solution to sprawl—as a means to develop compact, pedestrian- and transit-friendly areas in urban places that were previously neglected in favor of suburban development. For instance, a typical TOD objective is to have people drive less and ride public transit more (Cervero et al. 2004). This type of redevelopment is also often intended to attract new residents and consumers to the downtowns. For example, the San Francisco Bay Area Metropolitan Transit Commission (MTC) (Metropolitan Transportation Commission 2010) TOD briefing book describes the market which TODs are meant to serve: groups who prioritize access to high-quality transit and safe and convenient places to walk and bicycle, generally single or partnered professionals with no children.

Formed in 1993, the Congress for the New Urbanism (CNU), a group that includes architects, urban designers, planners, developers, bankers, historic preservationists, and government officials, committed to "traditional" American master-planned neighborhoods with minimal sprawl, began implementing an approach to planning common in White and economically prosperous neighborhoods (Saab 2001, Duany, Plater-Zyberk, and Speck 2000). CNU notes that Princeton, New Jersey and Georgetown in Washington, DC are prime examples of traditional towns. The CNU's charter, which it adopted in 2001, proclaims, "We dedicate ourselves to reclaiming our homes, blocks, streets, parks, neighborhoods, districts, towns, cities, regions, and environment (Congress for the New Urbanism 2001). Around the time the organization adopted the charter it formed task forces to deal with a range of planning objectives such as inner-city revitalization (Fainstein 2000, Leccese and McCormick 2000, Deitrick and Ellis 2004).

The Congress for New Urbanism references the resilience of home prices in "traditional" communities and the communities they hope to draw to the city center:

> No longer a secret, New Urbanism's advantages are reflected in the marketplace. Homes within convenient walking distance of stores, restaurants, and other amenities sell at a premium. They retain their values in up and down markets and are well positioned to accommodate growing ranks of aging baby boomers and small-household millennials.
>
> (Congress for the New Urbanism 2010)

The emphasis is urban design that creates dense, mixed-used, and pedestrian-friendly housing, with work-live lofts often the prime example (Talen 2005). New urbanists promote design that fosters security and community interaction, and respects local history, culture, and regional character (Fainstein 2000). The connection to TOD development is clear from the point that NU also calls for redevelopment that leverages a network of trains that connect transient-oriented neighborhoods to commercial and downtown districts.

New urbanists promote their vision by lobbying local governments to change traditional local government planning rules and adopt new policy mechanisms, specifically form-based codes (FBCs) (Katz 2008). Unlike traditional (Euclidean) zoning, FBCs codify a physical plan focusing on form and density. As defined by the Form-Based Codes Institute, an FBC is a "method of regulating development to achieve a specific urban form. Form-based codes create a predictable public realm primarily by controlling physical form, with a lesser focus on land use, through city or county regulations" (Form-Based Codes Institute 2008).

The new urbanism movement has been quite successful for new urbanists in terms of the amount of projects that have been developed or planned. In 2004, the New Urban News identified more than 600 new towns, villages, neighborhoods, and small urban infill projects in the United States that used NU principles (Steuteville 2004). In 2006 the CNU, U.S. Green Building Council, and the Natural Resources Defense Council launched over 200 NU city projects (Langdon 2007). As of 2015, there were over 4,000 NU projects planned or under construction in the United States, half of which are in historic urban centers (www.newurbanism.org/). These numbers support Talen's (2005) assertion that NU is the most influential American design movement of recent decades. The NU redevelopment model is indeed physically revamping cities for a new type of urban living.

Like TOD and NU, "creative city" or "creative class" redevelopment also calls on policy-makers and developers to target specific privileged groups. Though Charles Landry (2000) coined the concept creative city to indicate that city authorities should direct redevelopment resources that emphasizes creativity, culture, and design, Florida has expanded the ideas. Florida's (2005a, 2005b, 2002) "creative class," which I refer to in this book, includes a broad group of highly skilled artists, designers, technologists, producers, musicians, engineers, software developers, and business and education professionals who he postulates are essential to reinvigorate economic development growth.

Florida capitalized on his idea through his consulting work that garnered the attention of academics, city leaders, and economic developers, among others (Florida and Mellander 2014). Florida contends that this group has a preference for living and spending time in places that are diverse, tolerant, and open to new ideas. This includes spaces with cultural programming, art districts, technological development, and openness to ethnic, cultural, and sexual orientation diversity. The creative class, he adds, can benefit from central city real estate that features older buildings, mixed uses, and relatively low prices (Florida 2005a). According to Tom Low, director of town planning for Duany Plater-Zyberk and Co., the 25- to 48-year-old "creative class," which includes millennials, is "bored" with the suburbs and prefers city life (Metz 2005). Globally, Florida (Florida and Mellander 2014) found 11 out of 78 countries that he observed had anywhere from 40 percent to 47 percent of its workforce in the creative class. He also found that the United Sates had 35 percent of its workforce in the creative class.

Critique of Redevelopment Models

A growing body of literature questions the TOD, NU, and CC models in terms of their effects on disadvantaged groups and disinvested areas in central cities. For instance, some critics argue that TOD could lead to reductions in bus service and gentrify low-income neighborhoods (Hostovsky 2015). Detractors of NU argue that projects based on the model largely fail to provide an equitable mix of affordable housing and market-rate housing, and dislocate poverty instead of living up to the model's espoused principles of diversity (Vanderbeek and Irazabal 2007). Diaz (2012) argues that the private sector and city officials tend to support NU projects that help gentrify city centers and are economically out of reach for working families. Harvey's (1997) critique of NU gives us some ideas how this happens. He points out that the model is deterministic and advocates of the model prioritize capital accumulation and privileged groups. Wrongly cast aside by this NU movement, but crucial to have, he adds, are planning processes that are filled with "the struggle to advance a more socially just, politically emancipatory, and ecologically sane mix of spatio-temporal production processes." Likewise, some critics of the CC model have problems with its elitist approach to creativity and devaluation of the role of non-professionals in urban growth, and question whether the creative class can be the savior of the alleged blighted (Kratke 2012, Slater 2012). Appropriately, Brian (2005) makes a call for research that examines many of these conceptual critiques. An intention of this book is to help advance such research.

Racial/Ethnic Politics and the Grassroots Meet Urban Planning

Since the Civil Rights era, questions about racial equality have supported campaigns aimed at increasing the number of under-represented ethnic minorities in elected positions at all levels of U.S. government, with particular concentration at the local level where racial, ethnic, and class debates over redevelopment are especially visible. Community activists have advanced the notion that co-ethnic officials are more likely to understand and attend to the interests of racial and ethnic groups than members of other groups (Thomas 2008). At the policy level, for example, this logic presupposes that a co-ethnic elected official would approve the implementation of a redevelopment project or the allocation of resources to support the group's desired redevelopment policy, social program, or community development goals.

Community-Centered Approaches to Change

This section describes three types of community-centered approaches to change: community organizing, community coalitions, and participatory action research (PAR).

Community organizing has a strong tradition at the local grassroots level where racial, ethnic, and class debates over redevelopment are especially noticeable (Irazábal and Farhat 2008). Drawing substantially from American community organizer Saul Alinsky's (1971) conflict organizing dating to the late 1930s, community organizers employ tactics such as providing testimony in local government meetings, protesting, boycotting, staging marches and sit-ins, and circulating petitions (Gittell and Vidal 1998, Smock 2004). The Alinsky approach generally consists of an organizer who does not live in the particular area in question who works with local leaders of organizations and residents to openly identify and put pressure on a particular person they believe is responsible for and can solve a local problem. This requires the organizer to educate a broad base of organizations, such as religious groups, ethnic groups, political organizations, and labor unions about the politics that propelled the problem and involve them in actions in large numbers in order to confront and shame the "enemy." Community organizing of this type has a strong tradition in Mexican-American and Chicana/o neighborhood battles over affordable-housing, freeway construction, and other redevelopment policies (Acuña 2011, Davis 2000, Diaz 2005, Villa 2000). While the Alinksy approach remains alive and well in community development campaigns, there has also been newer and different community-centered approaches to change that are offshoots of the Alinksy model but de-emphasize conflict (Smock 2004).

Another example of community-centered approaches to change, community coalitions often consist of community-based organizations, residents, and other stakeholders coming together to deliberate, make decisions about, and take action to improve community needs and public policies (Baxamusa 2008). Coalitions typically identify and develop strategies to work with or against (or both) decision-making bodies that have the power to vote, implement, or in some manner influence the coalition's desired community development policy demands, such as city elected officials and community development staff and developers. Coalition work also often includes distributing leadership roles among members; leading workshops to educate members on urban planning technical terms and local politics; identifying community-wide allies who are not part of the coalition but could lend their endorsement to certain action plans; educating, learning, and recruiting from the broader community that they seek to represent; and networking to secure or increase resources to sustain the coalition (Roberts-Degennaro 1997, Hum 2010, Laing 2009). In recent years, some coalitions have crafted Community Benefit Agreements (CBA), forms of legal accountability from developers and local governments to communities in attempt to fulfill their developmental commitments (González et al. 2012). Some CBAs for community development have included items related to labor wages, affordable housing, and local urban governance (Gross, LeRoy, and Janis-Aparicio 2005).

PAR emphasizes collective inquiry about community problems with the collaboration of those affected by the problems being studied. It also entails education and reflection to understand what was learned and come up with tactics and activities that the group believes could effect change (Reason and Bradbury 2001, Minkler and Wallerstein 2003). The approach requires building skills among those participating, such that professionals in the collaboration provide members from the community with activities to reflect critically about what the main causes and potential solutions to local problems are, or what Freire calls "conscientization" (Lewin 1946, Freire 1970). A core idea of PAR is that collaborators should be trained in conducting research and use multiple research methods with the idea that this could yield a more comprehensive picture of problems. Some PAR community development projects have used surveys, focus groups, cognitive or risk mapping, photovoicing, town hall meetings, and asset mapping (González et al. 2007, Minkler and Wallerstein 2003, Israel et al. 2005, Silverman, Taylor Jr, and Crawford 2008, Reardon et al. 1993). The action component includes developing group empowerment whereby members gain confidence that collectively they are well prepared to successfully advocate or negotiate city-wide change. Learning Alliances is an evolved model of PAR for the purposes of community development in the UK which deeply involves public decision-makers in the knowledge coproduction processes together with communities and academics, increasing the likelihood of policy adoption and implementation (Moreno-Leguizamon et al. 2015).

About This Book

This book builds on what Wacquant (2008) calls *spatial alienation* to understand how the discourse and practice of redevelopment that intends to improve urban areas defines such areas as blight, further marginalizes and reduces to invisibility an already marginalized demographic, and seeks to drive the demographic out of sought-after areas for more desirable demographic groups (p. 241). Inherently, *spatial alienation* is about power. Yet, the ways in which spatial alienation discourse and practice manifests are not always obvious. This follows Foucauldian theory, which posits power as microphysics defining "innumerable points of confrontation" (Foucault 1977). That is, power is not always directly visible, such as unequal rights as embodied in disparate purchasing power in a housing market, but we can find it, if we look closely, embedded in all facets of civic and private life. Key to this book is uncovering the ways power manifests overtly or covertly in different forms and is represented in bits and pieces across different sources and material representations. Some ways power can be represented are words, images, and urban design, or when particular images, text, or urban design features are notably absent, highly decontextualized, or relegated to lower levels of importance.

This book has two goals. The first goal is to explore how city redevelopment plans and related city documents, as well as broader media and local publications, socially construct problems and imagined communities, erase existing and majority populations from redevelopment consideration, and create a set of practices prioritizing future communities. The second goal is to show how residents and grassroots leaders go about unbinding rather than accepting such plans, characterizations, and projects, the differences that some of these efforts make, and challenges that grassroots encounters along the way. The book emphasizes the feelings and attitudes that the spatially alienated and their activist partners expressed in the course of community activism against redevelopment that posed a threat to *La Cuatro*, and the downtown area more generally, as a culturally affirming and economically affordable zone in which to live and shop (Villa 2000, Wacquant 2008). It reveals how these groups responded to these threats, showing key victories and challenges that they encountered. My involvement in the most recent grassroots activities discussed in this book allows me to discuss them in a nuanced manner.

This book examines these goals against the backdrop of larger social, economic, and political environments that circulated at different time periods of this book's case study. This means that my analysis across the book's goals infuses racial, intra-ethnic, class, and privilege dimensions. I believe this book is the first to examine these topics in combination and do so in and interdisciplinary way with critical urban planning, ethnic studies, and urban sociology theoretical frameworks.

Research Methods

Latino City uses an in-depth case study with mixed and community-based methods. I conducted primary research from 2007 to 2012 as an active member of SACReD. My direct day-to-day involvement with the organization included participant observation, keeping meeting and field notes, and leading or co-moderating community-based action research processes. In the course of these activities, I used a qualitative approach to capture what I was observing in an interpretive, detailed, "thick" manner (Geertz 1973). This included participant observation (Yin 2003) and keeping meeting and field notes. My participant observation included attending Santa Ana Collaborative for Responsible Development (SACReD) general and subcommittee meetings, which collectively and typically amounted to about four meetings a month. On occasion, I also helped to prepare meeting agendas and facilitate meetings. I helped lead study sessions with the aim of interpreting the goals of the Renaissance Specific Plan (RP); assisted with creating coalition promotional material, disseminating the information about and recruiting for the coalition in the local neighborhood; and performed a leadership role in research and popular education activities, such as working with coalition members to co-develop, implement, analyze, and present data of a neighborhood survey and cognitive mapping exercises held during community forums. I also helped organize and attended community forums and other SACReD community-wide events. I attended city- and/or developer-sponsored community forums, city council meetings, and community benefits agreement negotiations with the city. My primary data collection continued past my regular involvement in SACReD. The ongoing changes to the businesses in the downtown area and gentrification debates that were taking hold in the area led me to conduct a survey of businesses in the downtown area in 2011 and this methodology is described in detail elsewhere (González and Guadiana 2013).

I reviewed SACReD archival documents that I maintained from the above activities, specifically from August 2007 to January 2012. This included my personal notes from general and subcommittee meetings, agendas from these meetings, and related documents, such as general field notes based on my involvement; the proposed CBA and all the primary data that were generated from the neighborhood survey and popular education activities to help create the CBA; the Station District plan; and city council minutes.

The main method for analyzing the SACReD process was process mapping (Huff and Jenkins 2002). Process mapping involves organizing project activities chronologically and then grouping activities by themes. Reflecting on the discussions and overall experiences that I had over the years as a participant-observer was instrumental during the construction of the process map. I worked closely with three SACReD members to refine the map, adding their recollections and personal material, and we categorized these recollections and sources of information and linked them to themes and sub-themes that emerged from the categories (González et al. 2012).

I also collected additional and secondary data that pertained to Santa Ana development from 1958 to 2015, using three techniques: critical review of archival documents, critical review of news/media documents, and critical review of marketing and promotional material. I reviewed over 40 city community development and related documents dating from 1958 to 1997 that are held in the Santa Ana Public Library History Room, including general plans and reports, redevelopment plans, community plans, consolidated plans, community development agency status reports, housing element plans and reports, planning commission annual reports, city-wide annual reports, and urban development federal grant applications and awards agreements, among other sources. I also reviewed over 50 city community development documents and related texts that I obtained either online, through public records requests, or through personal contacts. These included specific plans and Station District plans, city resolutions, city council meeting notes, tax credit reports, marketing and strategic plans of the city's public-private downtown business organization, grand jury reports, and police reports, all collectively ranging from 1985 to 2015. I reviewed over 120 local and regional Spanish language, English-Spanish bilingual, and English language newspapers that covered city redevelopment from the 1970s to 2015, accessing them through the Chicana and Chicano Resource Center at California State University, Fullerton, the California State University, Fullerton main library, online, or from my personal library. Other sources that I reviewed were blogs, news radio podcasts, posts on FaceBook, Instagram, and Twitter, and other online webpages. My data set also included, flyers, pamphlets, and other promotional and marketing materials for businesses and events in and around downtown and I collected these from time to time during my visits to downtown from 2006–2015.

Chapter Overview

Chapter 1 focuses on the formation of Santa Ana's *Latino City* identity from the 1960s and 1980s and how city officials attempted to comprehensively and rationally prepare for this change in the downtown. The emphasis is Mexican immigration and settlement in Santa Ana to meet regional low labor demand, city shift from rural to urban, and rapid demographic changes where Whites and Mexicans were similar in numbers. Racist, elitist, and rational planning discourse fostered spatial alienation while the city braced for these changes. English language newspapers and other local publications promoted the city's narrative. Collectively, the dominant discourse reduced or ignored nuanced cultural, social, and economic realities of place into a few melodramatic characterizations. The chapter also traces the birth of *La Cuatro* in the 1970s, the Mexican commercial zone that affirmed the heart of the downtown, and how the *La Cuatro* narrative contrasted the dominant discourse. Two questions guide this chapter: How do planners spell out their community development goals and plans to spatially alienate the central city downtown? What are some of the Latina/o community's views of and experiences with the central city downtown?

Chapter 2 shifts to a different era, the mid- and late 1980s, and reviews how the city seemed to commit to more democratic community redevelopment and downtown commerce interventions that reflected the city's growing Mexican families, immigrant base, and low-wage constituency. I examine these issues while showing how the institutionalization of proposed and actual disenfranchisement of the growing Latina/o community advanced, particularly through eminent domain, and especially in the downtown area. Focusing on two high profiled cases at the time, I ask: To what extent do struggles against eminent domain address a broader grassroots climate advocating for city reinvestment for the Latina/o community?

Chapter 3 looks at how the city transitioned to new urbanism, transit-oriented development, and creative-class approaches to redevelopment and began fulfilling the city's longstanding spatially alienating discourse and practice. Focusing on the 1990s and early 2000s, I trace how the discourse and practice of downtown redevelopment work to construct the imagined community along specific dimensions, erase the existing majority community from consideration, and build a set of practices to achieve these aims. The chapter focuses on two major redevelopment projects to examine these issues—the Artists Village and East End and the area's largest and busiest zone, *La Cuatro*. How do the discourse and practice of downtown redevelopment work to construct the imagined community? How does this erase the existing majority community from consideration? What are some practices to achieve these aims? Last, what does the downtown district look like during the transition from serving Mexican, working-class, immigrant consumers to serving CC and middle-class consumers?

Chapter 4 builds and expands on a previously published co-authored article on the (SACReD) (Gonzalez et al. 2012). The chapter explores why and how SACReD launched its own community-based planning approaches outside of formal public-private activities during TOD and NU project. Focusing on SACReD's pursuit of a legally binding agreement with the city and developer, the chapter seeks to answer two questions: How do community-based approaches address the needs and hopes of lower-income Latina/o urban areas undergoing TOD-NU? What are the challenges, contradictions, and limitations in practicing community-based participation?

The Conclusion offers final remarks which are an effort to synthesize what urban planning, politics, and the grassroots discussed in this book may suggest about the future of central city downtown redevelopment. I reiterate that the discursive dimensions of key city redevelopment plans, related city documents, and media and local publications have socially constructed problems, erased existing majority Latina/o populations, and framed understandings of desirable future communities. This discussion points out the changing discourse around racial and ethnic political representation, focusing on the key arguments made toward majority White city councils to the all-Latina/o city council. I argue that critiques of participation practices dating to the late 1960s continue to have much relevance in private-public participation practices that are commonly used to prepare contemporary new urbanist and transit-oriented development plans. The book makes the case that regardless of decade and revitalization strategy, redistribution of policy and resources for downtown commercial and housing options continued to favor the middle-class and native-born Whites, whether it terms them the creative-class, "artsy," or "hip." I offer directions for communities who wish to negotiate CBAs and fight displacement and gentrification, as well as questions to help guide future research examining these unjust practices across many cores of central cities in the United States.

References

Acuña, Rodolfo. 2011. *Occupied America: a history of Chicanos*. Boston, MA: Longman.

Alinsky, Saul David. 1971. *Rules for radicals: a practical primer for realistic radicals*. 1st ed. New York: Random House.

Baxamusa, M. H. 2008. "Empowering communities through deliberation: the model of community benefits agreements." *Journal of Planning Education and Research* 27:261–76.

Beard, A. Victoria, and S. Carolina Sarmiento. 2014. "Planning, public participation, and money politics in Santa Ana (CA)." *Journal of the American Planning Association* 80 (2):168–81.

Brenner, Neil, and Nik Theodore. 2002. "Cities and the geographies of' actually existing neoliberalism.'" *Antipode* 34 (3):349–79. doi: 10.1111/1467-8330.00246.

Brian D. 2005. "From good neighborhoods to sustainable cities: social science and the social agenda of the New Urbanism." *International Regional Science Review* **28** 217-238.

Cervero, R., S. Murphy, C. Ferrell, N. Goguts, Y. H. Tsai, G. B. Arrington, J. Boriski, J. Smith-Heimer, R. Golem, P. Peninger, E. Nakajima, E. Chui, R. Dunphy, M. Myers, S. McKay, and N. Witenstein. 2004. Transit-orinted development in the United States: experiences, challenges, and prospects. Washington, DC: Transportation Research Board. www.cnu.org/who-we-are/charter-new-urbanism. Accessed November 7, 2016.

Congress for the New Urbanism. 2001. Charter of the New Urbanism.

Congress for the New Urbanism. 2010 "Smart growth and new urbanism: it just performs better." www.slideshare.net/CongressfortheNewUrbanism/new-urbanism-just-performs better. Accessed February 4, 2016.

Davis, Mike. 2000. *Magical urbanism: Latinos reinvent the US city*. London; New York: Verso.

Deitrick, Sabina, and Cliff Ellis. 2004. "New urbanism in the inner city: a case study of Pittsburgh." *Journal of the American Planning Association* 70 (4):426–42.

Diaz, David R. 2005. *Barrio urbanism: Chicanos, planning, and American cities*. New York: Routledge.

Diaz, David R. 2012. "Barrios and planning ideology: the failure of suburbia and the dialects of new urbanism." In *Latino urbanism: the politics of planning, policy, and redevelopment*, edited by David R. Diaz and Rodolfo D. Torres, 21–46. New York: New York University Press.

Diaz, David R., and Rodolfo D. Torres. 2012. "Introduction." In *Latino urbanism: the politics of planning, policy and redevelopment*, edited by David R. Diaz and Rodolfo D. Torres, 1–20. New York: New York University Press.

Duany, Andres, Elizabeth Plater-Zyberk, and Jeff Speck. 2000. *Suburban nation: the rise of sprawl and the decline of the American Dream*. 1st ed. New York: North Point Press.

Fainstein, Susan S. 2000. "New directions in planning theory." *Urban Affairs Review* 35:451–78.

Florida, Richard L. 2002. *The rise of the creative class: and how it's transforming work, leisure, community and everyday life*. New York, NY: Basic Books.

Florida, Richard L. 2005a. *Cities and the creative class*. New York: Routledge.

Florida, Richard L. 2005b. *The flight of the creative class: the new global competition for talent*. 1st ed. New York: HarperBusiness.

Florida, Richard, and Charlotta Mellander. 2014. "The creative class goes global." In *The creative class goes global*, edited by Charlotta Mellander, Richard Florida, T. Bjorn Asheim and Meric Gertler, 1–7. Oxon, UK: Routledge.

Form-Based Codes Institute. 2008. "Form-based codes defined." www.formbasedcodes.org/what-are-form-based-codes. Accessed June 7, 2009.

Foucault, Michel. 1977. *Discipline and punish: the birth of the prison*. New York: Pantheon.

Freire, Paulo. 1970. *Pedagogy of the oppressed*. New York: Herder and Herder.

Frey, William. 2012. "Demographic reversal: cities thrive, suburbs sputter." www.brookings.edu/opinions/demographic-reversal-cities-thrive-suburbs-sputter/. Accessed January 16, 2016.

Geertz, Clifford. 1973. *The interpretation of cultures: selected essays*. New York: Basic Books.

Gittell, Ross J., and Avis Vidal. 1998. *Community organizing: building social capital as a development strategy*. Thousand Oaks, CA: Sage Publications.

Glaeser, Edward, and Jesse Shapiro. 2001. "Is There a new urbanism? The growth of U.S. cities in the 1990s." *NBER Working Paper No. 8357*.

González, Erualdo R., and Lorena Guadiana. 2013. "Culture oriented downtown revitalization or creative gentrification?" In *Companion to urban regeneration*, edited by Leary M. E. and J. McCarthy. Oxon, UK: Routledge.

González, Erualdo R., and Clara Irazábal. 2015. "Emerging issues in planning: ethno-racial intersections." http://dx.doi.org/10.1080/13549839.2015.1048975. doi: 10.1080/13549839.2015.1048975.

González, Erualdo R., Raul P. Lejano, Guadalupe Vidales, Ross F. Conner, Yuki Kidokoro, and Robert Cabrales. 2007. "Participatory action research for environmental health: encountering Freire in the urban barrio." *Journal of Urban Affairs* 29 (1), 77–100.

González, Erualdo R., Carolina Sarmiento, Ana Urza, and Susan Luévano. 2012. "The grassroots and new urbanism: A case from a Southern California Latino community." *Journal of Urbanism: International Research on Placemaking and Urban Sustainability* 5 (2–3):219–39.

González, Erualdo R., and Raul P. Lejano. 2009. "New urbanism and the barrio." *Environmental and Planning A* 41 (12):2946–63.

Grodach, Carl, and Anastasia Loukaitou-Sideris. 2007. "Cultural development strategies and urban revitalization: a survey of US cities." *International Journal of Cultural Policy* 13 (4): 349–70.

Gross, Julian, Greg LeRoy, and Madeline Janis-Aparicio. 2005. *Community benefits agreements: Making development projects accountable.* Washington, DC: Good Jobs First and the California Partnership for Working Families.

Harvey, D 1997. "The New Urbanism and the communitarian trap" Harvard Design Magazine Winter/Spring, number 1 http://www.harvarddesignmagazine.org/issues/1/the-new-urbanism-and-the-communitarian-trap

Harvey, David. 1989. "From managerialism to entrepreneurialism: the transformation in urban governance in late capitalism." *Human Geography, Geografiska Annaler B* 71 (1):3–17.

Hostovsky, Charles. 2015. "Transit oriented development without gentrification: towards smart cities that promote social justice." Accessed March 18. www.slideshare.net/EMBARQNetwork/transit-oriented-development-tod-without-gentrification-towards-smart-cities-that-promote-social-justice-charles-hostovsky.

Huff, Anne Sigismund, and Mark Jenkins. 2002. *Mapping strategic knowledge.* Thousand Oaks, CA: Sage.

Hum, Tarry. 2010. "Planning in neighborhoods with multiple publics: opportunities and challenges for community-based nonprofit organizations." *Journal of Planning Education and Research* 29 (4):461–77. doi: 10.1177/0739456X10368700.

Irazábal, Clara. 2012. "Beyond 'Latino new urbanism': advocating ethnurbanisms." *Journal of Urbanism* 5 (2):241. doi: 10.1080/17549175.2012.701817.

Irazábal, Clara, and Ramzi Farhat. 2008. "Latino communities in the United States: place-making in the pre-World War II, postwar, and contemporary city." *Journal of Planning Literature* 22 (3):207–28. doi: 10.1177/0885412207310146.

Israel, B. A., E. Eng, A. J. Schulz, and E. A. Parker. 2005. "Introduction to methods in community-based participatory research for health." In *Methods in community-based participatory research for health*, edited by B. A. Israel, E. Eng, A. J. Schulz, and E. A. Parker, 3–26. San Francisco, CA: Jossey-Bass.

Katz, Peter. 2008. "Introduction to form-based codes." Form Based Regulations: Creating Tomorrow's Communities Development Program presentation, Orange County American Planning Association Chapter, Fullerton, California.

Kratke, Stefan. 2012. "'Creative cities' and the rise of the dealer class." In *Cities for people, not for profit: critical urban theory and the right to the city*, edited by Neil Brenner, Peter Marcuse and Margit Mayer, 138–49. New York: Routledge.

Laing, Bonnie Young. 2009. "Organizing community and labor coalitions for community benefits agreements in African American Communities: ensuring successful partnerships." *Journal of Community Practice* 17 (1):120. doi: 10.1080/10705420902862124.

Landry, Charles. 2000. *The creative city: a toolkit for urban innovators*. London: Earthscan.

Langdon, P. 2007. "LEED aims to set 'First National Standard for Neighborhood Design'". *New Urban News*, September 27. www.newurbannews.com. Accessed May 2, 2009.

Lara, Jesus J. 2012. "Latino urbanism: placemaking in 21st-century American cities." *Journal of Urbanism*, 5 (2–3):95–100. http://dx.doi.org/10.1080/17549175.2012.693250..

Leccese, Michael, and Kathleen McCormick. 2000. *Charter of the new urbanism*. New York: McGraw-Hill.

Lejano, Raul P., and Erualdo R. González. 2016. "Sorting through differences." *Journal of Planning Education and Research*. Ahead of Print Articles. doi: 10.1177/0739456X16634167.

Lewin, Kurt. 1946. "Action research and minority problems." *Journal of Social Issues* 3:34–46.

Londoño, Johana. 2012. "Aesthetic belonging: the latinization and renewal of Union City, New Jersey." In *Latino urbanism:the politics of planning, policy, and redevelopment*, edited by David R. Diaz and Rodolfo D. Torres, 47–64. New York: New York University Press.

Main, Kelly, and Gerardo F Sandoval. 2015. "Placemaking in a translocal receiving community: the relevance of place to identity and agency." *Urban Studies* 52 (1):71–86. doi: 10.1177/0042098014522720.

Metropolitan Transportation Commission. 2010. Choosing where we live: attracting residents to transit-oriented neighborhoods in the San Franciso Bay area. A briefing book for planners and managers, edited by Metropolitan Transportation Commission. Oakland, CA.

Metz, Sylvain. 2005. "New urbanism: Attract the 'creative class' – but all is not in agreement." http://tidewatermusings.peterstinson.com/2005/12/new-urbanism-attract-creative-class.html. Accessed April 4, 2016.

Minkler, Meredith, and Nina Wallerstein. 2003. "Introduction to community based participatory research." In *Community based participatory research for health*, edited by Meredith Minkler and Nina Wallerstein, 3–26. San Francisco, CA: Jossey-Bass.

Moreno-Leguizamon, Carlos, Marcela Tovar-Restrepo, Clara Irazábal, and Christine Locke. 2015. "Learning alliance methodology: contributions and challenges for multicultural planning in health service provision. A case study in Kent, UK." *Planning Theory and Practice* 16 (1):79–96. http://dx.doi.org/10.1080/14649357.2014.990403.

Reardon, Ken, John Welsh, Brian Kreiswirth, and John Forester. 1993. "Participatory action research from the inside: community development practice in East St. Louis." *American Sociologist* 24 (1):69.

Reason, Peter, and Hilary Bradbury. 2001. "Introduction: inquiry and participation in search of a world worthy of human aspiration." In *Handbook of action research: participative inquiry and practice*, 1st ed., edited by Peter Reason and Hilary Bradbury. Thousand Oaks, CA: Sage.

Rios, Michael, Leonardo Vazquez, and Lucrezia Miranda, eds. 2012. *Diálogos: placemaking in Latino communities*. Milton Park, Abingdon; New York: Routledge.

Roberts-Degennaro, Maria. 1997. "Conceptual framework of coalitions in an organizational context." *Journal of Community Practice* 4 (1):91–107.

Saab, Joan A. 2001. "Historical Amnesia: New urbanism and the city of tomorrow." *Journal of Planning History* 6 (3):191–213.

Sandoval, Gerardo. 2009. *Immigrants and the revitalization of Los Angeles: development and change in MacArthur Park*. Amherst, NY: Cambria Press.

Silverman, Robert Mark, Henry L. Taylor Jr., and Christopher Crawford. 2008. "The role of citizen participation and action research principles in Main Street revitalization." *Action Research* 6 (1):69. doi: 10.1177/1476750307083725.

Slater, Tom. 2012. "Missing Marcuse: on gentrification and displacement." In *Cities for people, not for profit*, edited by Neil Brenner, Peter Marcuse, and Margit Mayer, 171–96. New York: Routledge.

Smock, Kristina. 2004. *Democracy in action: community organizing and urban change*. New York: Columbia University Press.
Steuteville, Robert. 2004. "The new urbanism: An alternative to modern, automobile-oriented planning and development." *New Urban News*, July 8. www.newurbannews.com/AboutNewUrbanism.html. Accessed April 22, 2009.
Talen, Emily. 2005. *New urbanism and American planning: the conflict of cultures. Planning, history, and the environment series*. New York: Routledge.
Thomas, June. 2008. "The minority-race planner in the quest for a just city." *Planning Theory* 7 (3):227–47.
U.S. Census Bureau. 2012a. 2010 "Census special reports, patterns of metropolitan and micropolitan population change: 2000 to 2010."
U.S. Census Bureau. 2012b. "Populations increasing in many downtowns, Census Bureau reports." www.census.gov/newsroom/releases/archives/2010_census/cb12-181.html.
Valle, Victor M., and Rodolfo D. Torres. 2000. *Latino metropolis*. Minneapolis, MN: University of Minnesota Press.
Vanderbeek, M., and C. Irazabal. 2007. "New urbanism as a new modernism movement: a comparative look at modernism and new urbanism." *Traditional Dwellings and Settlements Review* XIX (1):41–57.
Villa, Raúl H. 2000. *Barrio-logos: space and place in urban Chicano literature and culture*. 1st ed., *History, culture, and society series*. Austin, TX: University of Texas Press.
Wacquant, Loic. 2008. *Urban outcasts: towards a sociology of advanced marginality*. Cambridge, UK: Polity Press.
Yin, Robert K. 2003. *Case study research: design and methods*. 3rd ed., *Applied social research methods series*. Thousand Oaks, CA: Sage Publications.

1 Latino City Emerges, 1900–1980s

Studies of urban planning in the United States rarely mention downtown redevelopment of majority populated Latina/o, working-class, and immigrant central cities (Irazábal and Farhat 2008). The early emphasis of the central city redevelopment literature was describing how urban renewal has addressed the flight of middle- and upper-class White families to the suburbs, deterioration of older sections of the central downtown district, growth of working-class and poor communities of color in this area, and the factors that contributed to and resulted in the aggravation of economic and social problems.

Downtown central city redevelopment plans, decisions, and activity can overtly or covertly influence or be influenced by the larger social and political climates of the city, region, and United States. Racism and anti-immigrant sentiment, for example, can shape both the meaning and social value attributed to groups in the downtown central city and in turn, these meanings and social values can be institutionalized in redevelopment plans and related activity. This practice transcends the more "rational" approach to plan-making which suggests that planning is "objective" and "neutral" from political and other interests or perspectives. A critical analysis of redevelopment plans in downtown central cities undergoing or having a growing Latina/o, working-class, poor, and immigrant constituency is important because it may reveal a spatially alienating discourse.

To be sure, the community at large could also attribute their own meanings about groups and redevelopment to those that surface from redevelopment decisions, plans, and activities. Unlike state actors, the broader community could include any number of individuals, institutions, and community groups who may have greater flexibility to overtly or covertly express particular social narratives around redevelopment and in doing so directly or indirectly support or reject any one of them. The community at large could express narratives in any number of places, such as sharing public views at community and city hall meetings. Newspaper articles and other local publications could also express narratives.

Analyzed critically, narratives are not inherently isolated phenomena to any one institution and are best understood when tied to the larger social, economic, and political environments in which they are embedded. The nature and extent of narratives surrounding key redevelopment plans, decisions, and activity over time are a major focus of this chapter. I situate these processes to the early history of Mexican immigration and settlement to meet regional low labor demands, a feature common to the growth and locational patterns of Mexicans in the Southwest (Diaz 2005). While the White majority of Santa Ana and Orange County as a whole prospered economically during the early 1900s to the 1980s despite evolving economies, Mexican-Americans and Mexican immigrants generally did not fare as well. These patterns remained consistent while rural Santa Ana broadened its identity from the county's agricultural epicenter to an urban area known for manufacturing and other industries. As the 1970s unfolded, Santa Ana was rapidly growing in numbers, changing racially and ethnically, and becoming economically poor. Rapid demographic shifts where Whites and Mexicans were similar in numbers, overall population growth, and Santa Ana's shift from rural to urban were reshaping the city and city officials attempted to prepare for this change in the downtown. It is against these broader contexts that the chapter seeks to answer the following questions: how do planners spell out their community development goals and plans to spatially alienate the central city downtown? What are some of the Latina/o community's views of and experiences with the central city downtown?

Mexican Labor and Community, 1900–1970

Santa Ana is located in Southern California, about 30 miles south of Los Angeles, approximately 100 miles north of the United States–Mexico border, and 10 miles from upper-middle and upper-class coastal cities. Nestled in the middle of historically conservative Orange County, Santa Ana is bounded by the Garden Grove Freeway on the north, the San Diego Freeway on the west and south, and the Santa Ana and Newport Freeways on the east. The city has been the county seat since the late 1800s.

In the early to mid-1900s, the city had a majority White, U.S.-born, and relatively well-off population, but this has changed. Santa Ana Whites of native parentage constituted 71 percent of the population in 1890 (Haas 1995). Many of these migrants came from confederate states after the Civil War in pursuit of real estate ventures, land speculation, and other opportunities (Arellano 2008, Bricken 2002). The first major wave of Mexican immigration followed the 1910 Mexican Revolution, mimicking a Mexican immigration pattern occurring throughout much of the southwest region in that decade (Haas 1995). However, in 1939, Santa Ana included 32,000 residents, the overwhelming majority of whom remained White (Adams 1963).

Santa Ana was the epicenter of Orange County's booming economies well into the 1950s and Whites prospered. Santa Ana combined a decades-old agricultural industry with defense manufacturing, retail, and service. Marketing reports, such as the "Santa Ana California Market Analysis and Trade Study," show that White residents overall prospered quite nicely in these newer postwar economies, while Mexicans as a whole continued to work in the agriculture industry and did not fare as well (Dermengian 1955, Haas 1995). For instance, growth in the defense industry in the 1940s created thousands of jobs and ushered in a housing boom

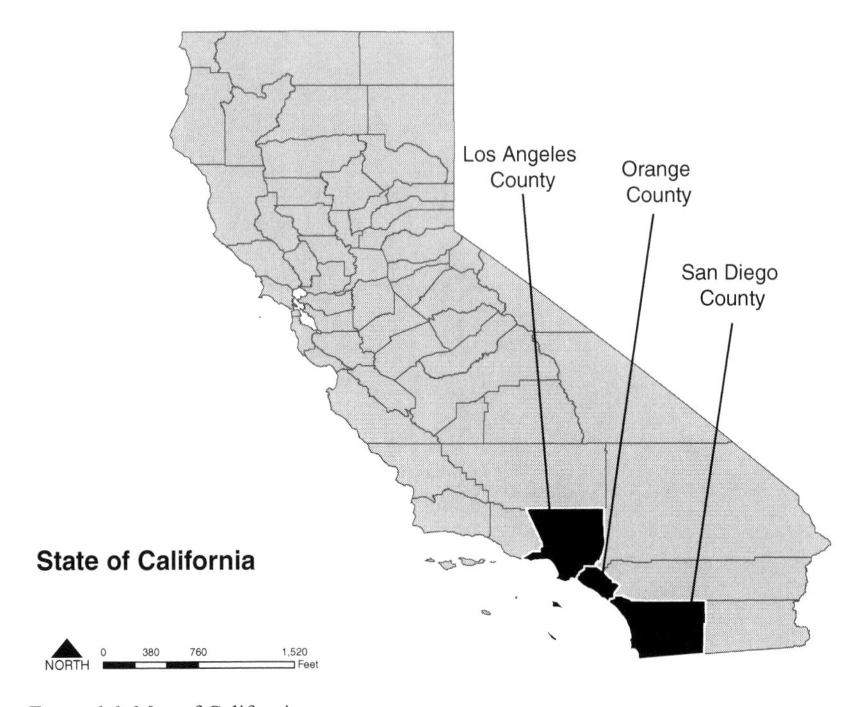

Figure 1.1 Map of California.

Source: Data from the U.S. Census Bureau; map created by Lorena Guadiana.

and an influx of military personnel and White-collar workers in Santa Ana and the surrounding region. Whites were the majority who benefited from the boom until around 1970 (City of Santa Ana 2009). While defense manufacturing had surpassed a still-thriving agriculture industry by the mid-1950s, wholesale and retail trade and service industry in the city grew in importance (Dermengian 1955, Santa Ana Planning Commission 1965). Regardless of work sector involvement, Santa Ana as a whole prospered quite nicely relative to Orange County. A market survey of a sample of Santa Ana residents in 1955 found that per-family income was about 5 percent higher than that of the rest of the county (Dermengian 1955). Economic prosperity certainly led to most of the population growth, and by 1960, the city population was 100,350, a post-1950 surge of 56,070, of whom Whites were over 80 percent (Ruth et al. 1965, Harwood and Myers 2002, Haas 1981).

Economic inequality surged between Whites and Mexican immigrants and the native-born of Mexican descent, or Chicanas/os, during these times. In 1960, both Mexican immigrants and Chicanas/os collectively were 15 percent of the population in Santa Ana, giving the city the largest Chicanas/os population in Orange County (Haas 1981). Yet Chicanas/os had been dominating the ranks of agricultural labor throughout the county during most of the first half of the twentieth century. Between 1914 and 1919, this agricultural labor force grew from 2,317

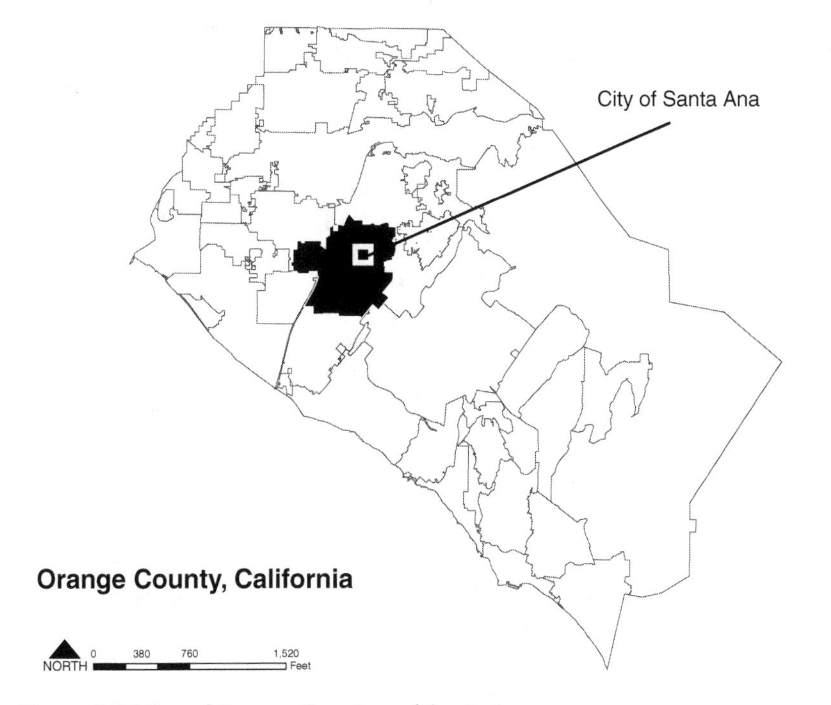

Figure 1.2 Map of Orange County and Santa Ana.

Source: Data from the U.S. Census Bureau; map created by Lorena Guadiana.

to 7,004 (Haas, 1981). Through the 1940s, this group had grown in dominance and composed an estimated 60 to 90 percent of the agricultural work force (Haas 1981). This group worked mainly in sugar and sugar beet factories, walnut groves, livestock farms, and vegetable farms.

In 1942 the United States and Mexico implemented a contracted guest worker program called the Bracero Program. This program, which formally lasted between 1942 and 1964, allowed agricultural employers in the United States to recruit and contract low-cost Mexican labor on a seasonal basis. For Orange County, the Bracero Program certainly helped boost the region's agricultural production and profits. Bank of America's "Focus on Orange County" report would term 1951–1960 the "golden years" in Orange County's agricultural production history. The agricultural sector contributed around $100 million to the county's economy in all but two years in this period (Haas 1995, p. 12).

Not all was golden, however, for Chicanas/os and Mexican immigrants in Santa Ana working in agriculture. The presence of the Bracero Program increased the underemployment and unemployment of a significant sector of the U.S. born Mexican descent community in the county, and the economic gains and quality of life of non-Bracero Mexican immigrants and Bracero immigrants were not on par with that of general Santa Ana residents (Haas 1995). Most White workers in Santa Ana continued to their employment in better-paying and newer industries.

The Formation of Latino City, 1965–1970

As the 1970s unfolded, Santa Ana city officials and residents did not need any census or related data to tell them what was obvious in and around the downtown and city neighborhoods generally. The city was changing dramatically; it was going from White and economically well-off to becoming *Latino City*.

With this change came scorned representations about and plans to eradicate this growth from key city stakeholders and community development plans from the 1960s to the 1980s. According to representations dominant at the time, Mexican immigration, settlement, and consumption, as well as businesses that marketed to these immigrants, resembled the working-class and poor "other" that was frighteningly transforming downtown life; this represented not rejuvenation of a commercial landscape blemished by patterns of building abandonment, neglect, and economic decline during the 1960s and early 1970s, but something radically different than the pre-1960s highly venerated middle-class, upscale, and White identity.

From the 1950s to1970s, the city's population began drastically growing and changing racially and economically. Santa Ana grew by 120 percent between 1950 and 1960. While Whites still dominated Santa Ana in 1960, their portion of the population would begin declining dramatically as they moved to new subdivision parts of the county and Latina/o immigrant populations grew (Harwood and Myers 2002, Goddard 1969). In Santa Ana, "the 1950s were the beginning of a changed outlook toward the coming industry, of unprecedented city growth, and a movement of churches, professional offices, businesses and shopping centers to the outer margins of the city and to the sites of new subdivisions. The mid-century period brought new dimensions to the city and a new environment for its citizens" (Goddard, 1969, p. 100). Santa Ana experienced the same White flight, or what this chapter calls "pulling," to the suburbs via racial housing policy patterns that were characteristic of many midcentury American cities. For example, FHA loans were more readily available in growing suburban areas across the United States than they were in central cities. And from 1934 to 1962, less than 2 percent of the $120 billion in housing financing underwritten by the government went to minorities (Mead 2003; Warner 2012). This type of federally backed racialized financing helped pull Santa Ana Whites to the suburbs. As an example of this suburban county growth, Orange County as a whole added 487,701 residents in the same decade from 1950 to 1960, a 225 percent increase, making it the fastest growing county in California (Adams 1963). In the next 18 years it grew 157 percent (Cannon 1978).

The socioeconomic level of the city also changed. In 1974 the estimated median family income of families in Santa Ana was $10,110 compared to $12,981 for the entire county (Urban Futures, Inc. and Planning Department City of Santa Ana 1975). According to the 1976 Special Census, Santa Ana residents were more heavily represented in lower skilled, lower paid jobs (e.g., craftsman, operatives, laborers, service workers) compared to the whole county (The Arroyo Group 1981).

Housing valuation and patterns also altered. Santa Ana was one of five cities in the county whose proportion of the county's highest-valued housing units declined significantly from 1960 to 1974 (Urban Futures, Inc. and Planning Department City of Santa Ana 1975). Meanwhile, countywide, dwelling units expanded from 79,158 in 1950 to 550,576 in 1973, and most of the homes in new cities were being added at the expensive and middle-valued range (Urban Futures, Inc. and Planning Department City of Santa Ana 1975). By 1974, for example, the cities of Laguna Beach, Newport Beach, Seal Beach, Villa Park, Irvine, and unincorporated Tustin-Foothills held a disproportionately large share of the most expensive homes with regard to their percentage share of the county's total housing stock (Urban Futures and Planning Department City of Santa Ana 1975). While homeownership flourished across the county, different housing patterns were emerging in Santa Ana. The city now had more renters than homeowners from 1970 to 1974.

The city also witnessed the emergence of *Latino City* during the 1960s and 1980s, the first period of rapidly changing, browning, and immigrant demography (see Figure 1.3).

This sharp change happened between the 1970s and 1980s when Latinas/os accounted for the largest population growth in Santa Ana. For instance, in 1970, the Latina/o population was 40,000. The population began growing at a steady rate of 4.6 percent per year and by the mid-1970s the boom was much more substantial, accelerating to over 19 percent per year towards the end of the decade, now representing over 90,000 people (The Arroyo Group, 1981). Much of this demographic growth stemmed from immigration. Whereas the city's immigrant population was just under 7 percent (6,959) in 1960, by 1980 immigrants were 30.5 percent (63,159) of the city's population, the vast majority from Mexico. This means that there was an almost 800 percent immigration surge across two decades. Not surprisingly, the city entered the 1980s having as many Latinas/os as Whites; Whites constituted 45.4 percent (97,633) and Latinas/os 44.5 percent (95, 697). Asian and Pacific Islanders (many refugees) and Blacks combined for 9 percent (10,322), and American Indian, Eskimo and Aleut, and "Other" groups accounted for less than 1 percent (U.S. Census Bureau 1980).

Clearly, the city at large was rapidly changing racially, ethnically, and socially from White and middle class to Mexican immigrant and working class. Where once a handful of barrios throughout the city and a few affordable consumption spaces in the downtown area's east portion marketing to Mexican-Americans and Mexican immigrants were spatially segregated from majority White residential and consumption areas, they had now, by the mid-1980s, transformed rapidly within a decade or so to be commonplace in that Mexican immigrant working-class families were in almost every neighborhood and the downtown's historically busiest, bourgeois, and White commercial street, Fourth Street, now had a Mexican, working-class, and immigrant character.

Becoming more common in the city too was the everyday reference by Spanish-speaking locals to the city as Santana (instead of Santa Ana), an uninterrupted segment of speech with three syllables combining two words into one, and to Fourth Street as *La Calle Cuatro*, or *La Cuatro* for short, which

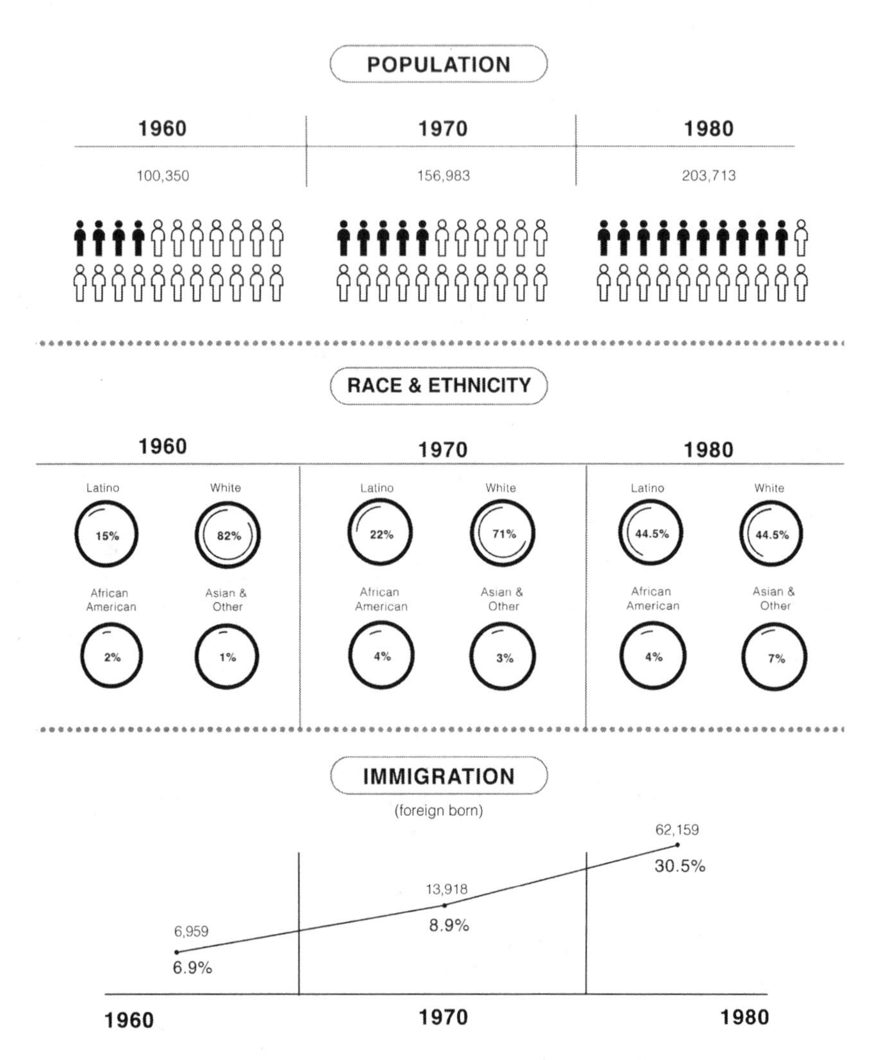

Figure 1.3 City of Santa Ana population by race/ethnicity and immigration, 1960–1980.

Source: Data from the U.S. Census, 1960. Table 32–Summary of Social Characteristics, For Standard Metropolitan Statistical Areas, Urbanized Areas, and Urban Places of 100,000 or More: 1960; SOCDS Census Data Output: Santa Ana City, CA; The Arroyo Group, ERA, PBQ and D, POD, Inc., and Melvyn Green and Associates. 1981; Santa Ana General Plan Revision Program: Problems and Opportunities Report. Figure created by Cara Gafner.

is the Spanish-language translation of the street. Latinas/os, the vast majority of immigrants, and first-generation residents of Mexican descent were now a major segment in the city and well on their way to shaping what was rapidly becoming *Latino City.*

Shifts in population were reshaping the city. Santa Ana had cemented its Latino City identity and Latinas/os would go on to sustain a consumer interest in downtown, even as income levels in the city stagnated. City officials attempted to prepare for this change in the downtown.

The Birth of Spatial Alienation, 1965–1980

The city was slowly becoming less economically well off and White and it had a great desire to keep up with newer cities' commercial success by the mid-1960s. These trends routinely preoccupied city officials when they talked about and planned for returning downtown to the pre-1960s "glory days." Figure 1.4 represents this downtown "fabulous fifties" time period in 1955 (Goddard 1969, p. 92).

By the mid-1960s, the city began the incipient origin of spatial alienation discourse and practice for the downtown (Figure 1.5). The discourse begins with the city's new phase in city planning that set the stage for decades of activity to de-Mexicanize, de-immigrant, de-concentrate poverty, and re-Whiten the downtown area and make it more upscale. Wealthy White consumers were the city's desirable consumer in the downtown. Newer, White, well-off cities in Orange County had a better tax base than Santa Ana that the city wished to attract. The city's 1965 Proposed General Plan for Santa Ana is a case in point. Yet, while it avoids any mention of racial, ethnic, or class urban trends, its concerns are obvious:

Figure 1.4 West Fourth Street, circa 1955.

Source: Courtesy of the Santa Ana History Room/Santa Ana Public Library.

Figure 1.5 West Fourth Street, circa 1965.

Source: Courtesy of the Santa Ana History Room/Santa Ana Public Library.

> The increasing population and expanding area of the City changed the nature of local planning. Before the postwar spread of urbanization, planning decisions in Santa Ana were relatively infrequent and simple.
> Planning in Santa Ana has since become more complex. Decisions must be made frequently and must be based on extensive knowledge, not only of the community, but also of regional trends and plans which might affect the community.
>
> The rate of change, the size of private and public investments, and the new complexities of a more urban environment have all pointed the need for a new general plan. (Santa Ana Planning Commission 1965, pp. 4–5)

The reoccurring emphasis in city community development reports and commissioned studies at this time was that Downtown Santa Ana had lost businesses to newer shopping centers and malls within the city and in Orange County generally and as a consequence Fourth Street had emptied out or become blighted. Within the city, local and major mid-range and higher-end stores had relocated to newer strip and shopping malls, such as Honer Plaza and Fashion Square, both of which opened in the late 1950s. While the city consistently reported that its major retail centers outside of the downtown were generating moderate to good sales tax revenues, these revenues were below those of similar or larger retail centers in other cities. This was a major problem to the city.

The city routinely conducted or commissioned reports to express and justify concern. The city's assessments of the downtown sales tax revenue during the 1960s and 1970s almost always involved brief qualitative assessments of an underperforming zone, accompanied with mentions of blight. For instance, downtown Santa Ana had more competition after World War II, both within the city and among newer cities in the region. While a few major shopping centers built in the early 1950s and 1970s outside of downtown were doing quite favorably overall, the city's per capita retail sales from 1970 to 1980 were lagging behind those of Orange County as a whole (The Arroyo Group 1981). This worried the city greatly. Orange County's overall taxable apparel sales increased 20.2 percent from 1976 to 1979, while Santa Ana's apparel sales increased only 7.4 percent per year in the same period (The Arroyo Group 1981). Santa Ana's regional retail competition included several new shopping malls that sprang up in Orange County but still very close to Santa Ana, such as the luxurious South Coast Plaza in Costa Mesa, which opened in 1967 bordering Santa Ana city limits and eventually became routinely, even to this day, the highest grossing retail shopping mall in the United States (Canalis, Zint, and Miller 2015, Fashion News 2015). The city was also alarmed by the low purchasing power of city residents relative to county-wide residents. In 1979, residents of Santa Ana had 25 percent less purchasing power than the general Orange County population ($17,854 vs. $22,350) (The Arroyo Group 1981).

It is unclear how well downtown Fourth Street fared economically relative to other retail zones in the city and across the county during the late 1960s and 1980s, however. Santa Ana did not report downtown Fourth Street area sales tax revenues in its general urban plans and community development reports in any consistent manner. Community development plans and related reports do not provide taxable retail sales data for Fourth Street for any year from 1965 to 1980. This makes unclear what quantitative comparisons to other centers within and outside the city would yield. As a whole, nevertheless, a consistent message in all these plans and reports was that Santa Ana, especially downtown Fourth Street, was not doing well economically relative to Orange County.

The city entered the 1970s armed with worrisome economic data and for the first time constructed a discourse of urban degeneration and call for a future with White and economically well-off patrons. For city officials, Downtown Fourth Street and the downtown in general were in a major urban crisis and they needed to redevelop the core immediately. The consistent message was that the city needed to act quickly to revive the once glorious downtown because it was blighted, had been doing economically very poorly, and had been socially undesirable for way too long. The discourse and practice of spatial alienation that evolved over urban space were comprehensive and the basis to erase or minimize distinctive and affirmative cultural and working-class qualities of the downtown commercial character and local neighborhoods that were on the rise.

Equipped with new community development powers in the early 1970s, state actors embarked on an experimental agenda to frame a planning discourse that emphasized that the downtown area was heavily physically distressed, structurally hazardous, and economically underperforming. This multiplex rhetoric contained an integrated social-cultural component as well, one that painted the downtown as socially dangerous and culturally marginal. Likewise, this narrative set the stage for the city to re-invent an ideal community and place that could replace the existing one. The city was banking on the possibility of middle-class and White renewal.

State actors engaged their representations of place to legitimate and facilitate their community redevelopment actions. In December 1972, the City Council declared that the city desperately needed to revitalize the central downtown core (Community Redevelopment Agency 1973). The city wasted no time, and a month later in 1973 it had formed a Community Redevelopment Agency to assume a lot of the revitalization responsibilities, as required by recently passed California law. The law had shifted new planning power and responsibility to cities, and Santa Ana now had authority to carry out new redevelopment activities, such as creating a renewal agency, surveying the central city to determine redevelopment project boundaries, and, based on discovered "blight" and "substandard" structures in these boundaries, designating and planning renewal projects, acquiring and disposing of real estate, and financing, all without raising taxes (Kennard, Delahousie, and Gault 1974; Community Redevelopment Agency 1973). Within six months, the city drafted its own preliminary downtown redevelopment plan and submitted it to the Redevelopment Agency/City Council for approval. The city established geographic boundaries for redevelopment and declared downtown blighted (Snow 1987). The plan's redevelopment objectives focused heavily around image, historic preservation, and economic development:

> To restore the economic and social health of the Santa Ana downtown area, make the area a source of pride to persons residing and working in Santa Ana or visiting the City, guide development towards an urban environment preserving the aesthetic and cultural qualities of the City, assist in the re-establishment of businesses within the project area, and stimulate and attract private investment thereby improving the City's economic health, employment opportunities and the tax base. (Community Redevelopment Agency, p. 3)

While image, historic preservation, and economic development were the city's concern in this preliminary plan, the city subsequently and immediately made it clear that it was contending with more specific concerns—downtown buildings that had been neglected by their owners, loss of high-end businesses, and rapid demographic change. The plight of the downtown, however, not only required addressing these concerns, but also necessitated characterization of "social blight" to justify redevelopment interventions.

Throughout the 1970s, waves of immigrants settled in Santa Ana, ushering in the city's first major legion of immigrants from Latin American, virtually all from Mexico. The number of immigrants surged from about 14,000 in 1970 to over 62,000 in 1980. This immigrant growth helped push the Latina/o population from 22 percent to 44.5 percent during the same period (see Figure 1.3).

The immigrant population generally resided throughout the city, including historic barrios in the downtown area, such as the Logan Barrio and Civic Center Barrio. These barrios had largely been home to earlier generations of segregated Mexican immigrants and their Mexican-American children (Garcia 2007). These barrios and other working-class and poor and ethnic minority neighborhoods in the downtown area stretched mainly east, south, and west within the downtown core.

Nestled among and around these barrios and neighborhoods were a few middle-class and White neighborhoods that collectively served as a symbol of power to many in city. These downtown neighborhoods mainly consisted of craftsman homes, and adjacent north of the downtown core was the city's most affluent upper-middle-class neighborhood adorned with an eclectic mix of large vintage homes dating to the 1920s. These more affluent neighborhoods were a symbol of power in the city because that is where the majority of elected city officials had historically resided. These working-class and ethnic minority neighborhoods and barrios and more affluent and White areas were starkly contrasting worlds of the status quo and "other."

Despite the clear disparity in material conditions between these working-class and poor neighborhoods and the better-off neighborhood classes, city redevelopment plans and mainstream commentary on the lifestyles and behavior of the poorest immigrants and ethnic minorities was little concerned with issues of living costs and more with what were perceived as consequences of the inadequate qualities, immoral practices, and deficient purchasing power of the poor. As Diaz notes of this general time period in the Southwest, which applies to Santa Ana, natives of Mexican descent (Chicanas/os) and Mexican immigrant working-class households "were trapped between two urban policy vises: racism and high living costs" (p. 86).

Given the social and economic characteristics ascribed to its growing and majority population, Fourth Street and working-class and ethnic minority barrios and neighborhoods figured predominantly in the 1974 *Downtown Santa Ana Development Plan* (DSADP; Kennard, Delahousie and Gault 1974). This plan was at the time the city's most comprehensive downtown plan. The plan sought to represent a more systematic and "rational" approach to the city's preliminary redevelopment plan written less than a year prior.

A critical look at the plan, however, reveals it was not divorced from constructing a community in the margins and reimagining an all-together new one that could resemble the downtown area's last standing middle-class White neighborhoods. The city took some steps to help ensure that the architecture firm consultants who prepared the plan met the objectives from the city's preliminary plan. The DSADP was essential to the city because it would supplement the existing redevelopment plan with now-required specific land use recommendations, such as the amount, density, and location of potential land uses (p. 1). Equally important, the DSADP would conform to the redevelopment objectives and policies of the newly created Community Redevelopment Agency (p. 1). For instance, the DSADP would provide the city with language and definitions that were recently and officially recognized in federal and state urban planning circles, such as "blight." The plan would also provide related data and assessments that the city did not thoroughly account for in earlier plans, such as an evaluation of the extent to which Santa Ana could be successful in becoming Orange County's downtown commercial center of attraction in light of recent physical, social, demographic, and economic changes.

According to the DSADP, the plan was a comprehensive and rational community development tool to guide the city into the future. The plan pointed out that it was prepared through interviews with relevant city departments and personnel, developers who were interested in investing in the area, field observations of the downtown area, and reviews of existing data and other sources, such as economic reports. The resulting DSADP was a document with three plan options that the city had an opportunity to choose from, a discussion over which plan option the city chose and why, and the consultants' recommendations based on the city's plan choice. The city had, for the first time, a viable opportunity to articulate its spatial alienating discourse and propose large-scale redevelopment activity.

The Discourse of Blight

The DSADP would be the first plan where the city expressed and legitimized prevailing discourse that downtown dwellers, consumers, and general visitors were driven into rampantly available criminal, foreign, and lower-class cultures, among others in a list of defamatory descriptions, an image condition that would only worsen and preclude all aspects of a White and middle-class defined population without government redevelopment intervention. The plan would also make clear that efforts to create new uses for residential and commercial spaces for a White and middle-class community invariably required omitting any substantive plans for existing majority residents and consumers, namely working-class and ethnic minorities.

The plan socially constructed problems and imagined communities and erased the majority population from redevelopment consideration. From the start, the DSADP made it clear that a series of major obstacles combined to "isolate the downtown" and "the net result has been a weakening of downtown market demand" (DSADP 1974, pp. 4–5). The plan listed such obstacles as Santa Ana's new and growing population and the social environment. The city did not take these concerns lightly. For example, the DSADP stated that during the preparation of the plan, key city departments had consistently expressed that "without question the social problems related to population transiency and crime must be alleviated and a stable downtown population created to facilitate successful redevelopment" (p. 26). Certainly, the plan listed additional obstacles such as the tremendous deterioration of buildings; road, traffic, and transportation issues; growing housing and commercial activity in the outskirts of Santa Ana and suburbanizing Orange County; and the relocation of businesses and industry from Santa Ana to other cities. My focus in turn, however, is to analyze the plan's dominant discourse around populations.

The DSADP's "Social Profile" map (see Figure 1.6) is a defamatory characterization of the downtown's population.

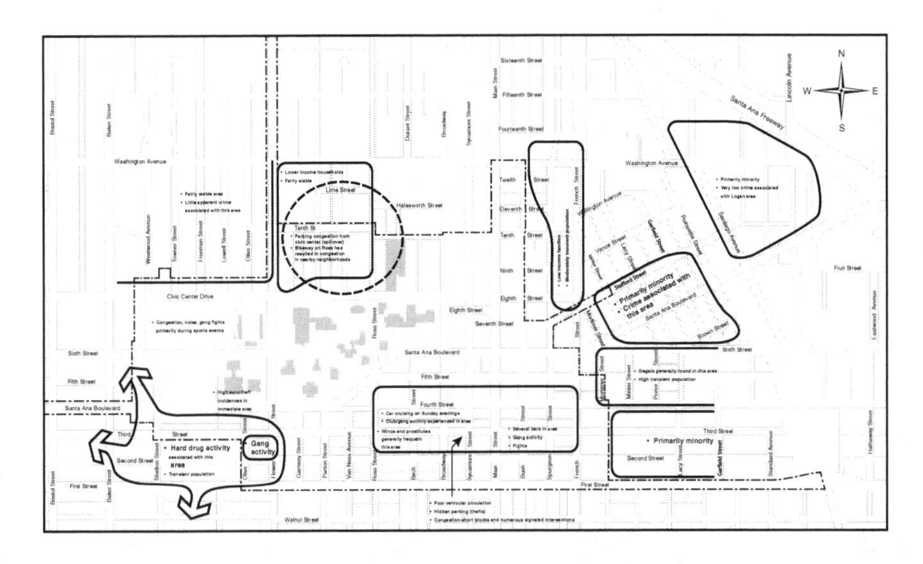

Figure 1.6 Downtown Santa Ana Development Plan "Social Profile."

Source: Courtesy of the Santa Ana History Room/Santa Ana Public Library. Figure created by Kelly Donovan.

In part, the social profile is designed to express a threat to safety, which in turn indirectly signaled that this would inhibit the desires of potential new entrants to live, consume, or visit the area. The map depicts where these dangerous populations were concentrated, with slight qualitative and no quantitative detail associated with crime and safety. In doing so, the map vilified the area and local community. For instance, the main problems along Fourth Street were described as "car cruising on Sunday evenings … club/gang activity … winos and prostitutes generally frequent[ing] this area … [and]…several bars in [the] area" (p. 27). Criticisms of the downtown neighborhoods, particularly on the east and northeast of Fourth Street, included "illegals generally found in this area … high transient population … and … primarily minority … [and]… crime associated with this area" (p. 27). This Social Profile conceives a particular alienating image of space (Lefebvre, 1991). These characterizations suggested that on any given day or evening, crime, criminals, or socially undesirable activities or people were overrunning downtown. The juxtaposition of "illegals" and "minorities" in the social profile squarely implicate Mexican-Americans and Mexican immigrants to this downtown scene, all of which are the antithesis of images of what a middle-class community in a revitalized area proposed by the city would look like. With this image of space, the map and profile are certainly indicative of plans for revanchist gentrification (Smith 1996).

The DSADP also pointed out that building blight and the social profile combined to produce unacceptable changes in merchandising and marketing methods (p. 10). Lamentably for the city, there was an abundance of stores that had declined in the price and quality of goods and services they offered, "a lot" of outdated and unattractive storefronts and interiors relative to newer shopping centers, and an increase in the number of vacant stores and secondhand businesses.

According to the DSADP, dated and unappealing commercial buildings stemmed from the "deferral" of routine maintenance and improvements to buildings. The plan avoided using the word *neglect* and did not implicate any responsible stakeholder in this neglect, such as building owners, real estate investors, or city departments (p. 10). All the dated deteriorating buildings were only part of the problem. The existing businesses were also problematic because they were low-cost stores and targeting the population that resided next to or within the downtown. This occurred with the DSADP casually noting that downtown employees and visitors of government offices are also patrons.

The DSADP reported that the city had also, unintentionally, contributed to Fourth Street's decline. The DSADP noted that patrons from other cities were no longer coming to downtown because the city's recently constructed Civic Center Complex had inadvertently shifted perception of downtown's commercial identity among general Orange County residents to primarily government and financing. The plan stated that retail activities were now serving a secondary function and in addition, were primarily serving the local population. Based on this logic, then, the regional market (code for White middle-class residents of Orange County) presumably were under the impression that the city was no longer interested in having a retail center for them and hence, this regional market ceased to frequent the downtown. Accordingly, the historically White middle- and upper-middle-class customer base had fled and the city now referred to it with the "regional" term and "local" became synonymous with "minority," "illegal" and "low-income" communities. Collectively, these "problems" were stunting downtown revival. According to the plan, downtown now lacked focus, missing particular types of businesses and a desirable demographic. A coherent focus required primarily consumption and amenity options for the middle-class White community.

As a corollary to "blight," the DSADP erased the existing and majority population from redevelopment considerations. The DSADP's final assessment[1] of the extent to which the city could fulfill the regional goal erases mention of families and residents living in and around the downtown and privileges civic professionals to reap from redevelopment. This comes with the plan's mention that there was very little hope for the city to create a regional commercial retail core. Not all was lost, however, and the plan still provided the city some hope to capture part of the demographic it was yearning for, though in a less intense fashion. Instead, the DSADP recommended that the city's overall redevelopment effort focus on serving the needs of populations who were guaranteed to be in the area on a regular basis: professionals of and visitors to government and financial enterprises; again, the local resident community was not included in this recommendation (p. 55). While the DSADP did not provide numbers regarding where civic employees lived, the plan stated:

> Discussions with potential private developers and analysis of the financial costs and benefits indicated that it would be extremely difficult to develop a regional shopping center within the core area. In contrast to a regional center, a community center would be more logical for the downtown and could service the market provided by employees and visitors to the government,

> financial, and professional activities. With a community center servicing the downtown market, we examined potential locations for a regional type center and found that project expansion to a site north of the Santa Ana Freeway with good access and visibility could be appropriate for a regional center. (p. 55)

The Introduction of Mexican Urban Design

While the DSADP erased the Mexican-American and Mexican immigrant population, it recommended that redevelopment incorporate Mexican urban design as part of the broader plans for the downtown core. In terms of urban design and land use, the plan recommended a commercial retail core with professional offices and within it a "Mexican village" theme along Fourth Street (pp. 38, 55, 66). The "Mexican village" would presumably appeal to city employees.

The core would have overlapping components totaling over 200,000 square feet of low-density and low-rise structures with a specialty retail zone and office complex and within it, a superblock on the west side consisting of "Mexican village" themed retail, offices, and a cultural area. This theme project would include additional movie theaters, retail, restaurants, an auditorium, and parking and apparently complement some of the area's historically and architecturally significant buildings (requiring rehabilitation) that according to the plan would lend itself to the "Mexican village" concept.

The DSADP shows that the city did not approve this alternative recommendation because it claimed it was not economically feasible. The consultants were rather insistent on their recommendation, however, and recommended that the city not give up on the idea and at the very least implement either of the two themed projects at a smaller scale (p. 55). The plan stated:

> The feasibility of a combined retail and professional office theme project in an "old town" or Mexican village" style was evaluated. The existing development proposal was not financially favorable to the Agency due to its proposed density and floor area ratio because the revenue generated would not offset the Agency costs. We recommend that the theme type use be included in the Final Development Plan, however with less land area and higher intensities of activity than originally proposed. (p. 55)

These urban design recommendations are revealing. First, the "Mexican village" theme "embraced" Mexican culture as an urban design concept, yet erased the local Mexican community consumer and resident as a beneficiary from any community development and related consumption options that would come from it. The embracing comes through suggesting Mexican-themed urban design and land use recommendations, though the plan stopped short of offering urban design details or reasons why the consultants made the recommendation and erased the local community. Absent in the plan is any discussion that suggests how the themed project could address the cultural, economic, or other needs and aspirations of the local and growing consumer and residential population. This erasing makes it clear that the DSADP and the city were not ready for redevelopment supporting and maintaining downtown spaces that were beginning to emerge for poor, working-class, and immigrant populations.

The plan also gave an additional meaning to "erasing," which was the recommendation to condemn housing. Specifically, this meant condemning Civic Center Barrio, proposing to erase the community from actual existence in the area. The plan pointed out that the city should rededicate existing land with housing to new housing construction, though the plan did not specifically name particular neighborhoods or barrios. The plan stated that residential single family housing, which included duplexes and rooming houses, concentrated just west of the proposed themed project, were an obstacle to overall redevelopment. According to the plan, the problem was that the current area with housing "has a high number of substandard structures" (p. 83). The DSADP recommended clearance as "infill," since "land and development costs prohibit scattered site housing and rehabilitation in this area" (p. 83). In other words, the plan recommendation was to tear down the existing Civic Center Barrio.

Plans for demolishment did not end there, however, and Fourth Street would be a prime candidate. The city had larger schemes to condemn or renovate much of the downtown area. This activity could be justified or facilitated with the plan's data pointing out that blight existed in almost every downtown block. For instance, the plan outlined that almost all of the downtown's commercial stock across 320 acres were uninhabitable or dangerous and required significant redevelopment activity. The plan reported that 67 percent of the commercial buildings were "substandard,"[2] the lowest rating possible; the remaining 23 percent had seismic deficiencies that required upgrades, and many of these buildings were more than 30 years old (p. 77). The DSADP findings that 90 percent of the downtown area required major rehabilitation and/or clearance certainly provided city officials and hopeful developers data to support the need for ambitious redevelopment. Fourth Street was a clear candidate for such attention. Indeed, while the DSADP reported that although most commercial activity in downtown was still concentrated along Fourth Street, the cumulative delays in upkeep to many of the buildings led to "physical and economic deterioration" and would now require a lot of rehabilitation and/or clearance (pp. 10, 77).

Imagining a New Community

A final plan component required defining the population that the city desired. Santa Ana and Orange County had developed tremendously in recent years, and ethnic minorities and the working class were increasingly living and shopping in the downtown area. City leaders realized that they could not rest on Santa Ana's pre-1960s laurels to attract White middle-class and upper-class people. Officials were eager to plan aggressively. For instance, this included defining a desired target population, which was White middle- and upper-class Santa Anans who had stopped consuming in downtown, either because they had moved to a different city or remained in the city but ceased to frequent downtown. While the DSADP and related city documents do not explicitly or consistently provide data on these trends, the city routinely reported in these documents that a significant block of that demographic had moved to regional cities and was shopping elsewhere. This may have been the case; 13 cities in Orange County incorporated during the 1950s and 1960s (Orange County Historical Society 2015). The target population included White middle- and upper-class residents from general Orange County.

While the DSADP as a whole established a clear spatially alienating narrative, it was also blunt that the city would very likely encounter tremendous obstacles in pursuit of attracting the desirable demographic groups. The question confronting the city, stated the DSADP, was whether it could actually address the obstacles described in the plan and in other reports that were making it difficult for the city to return to the golden years. While the DSADP recommended that the city immediately focus on eliminating physical blight, it also recommended that the city decide how much development must occur to create a change in the visual and general perception of downtown (pp. 70, 80). The tone to this recommendation suggests that the plan's consultants were cautioning against accelerated displacement and gentrification.

These and similar decisions would continue to preoccupy city officials. One such example is the city's 1979 Housing and Community Development Block Grant application to the U.S. Department of Housing and Urban Development, which shows this grave concern and uneasiness. City officials succinctly and clearly declared their racist and classist views towards Santa Ana's growing population: "The City of Santa Ana … contains some of the worst … social blight in the Southern California area. Santa Ana houses the largest minority population, estimated at 36 percent of a total population of 183,900 not including an estimated 60,000 illegal aliens"(City of Santa Ana 1982). "Social blight" was one in the same with the "minority population" and "illegal aliens." The city was also dissatisfied with affordable housing that provided homes for this community. In the city's 1981 *Problems and Opportunities* community development report, city officials frankly stated their disdain with Santa Ana's low-cost housing options relative to a more elite county housing stock, which demographic trends suggests were taken up mostly taken up by Mexican immigrants and the working class: "Santa Ana is sometimes referred to as the 'houser of last resort,' providing as it does, a large portion of what low cost housing exists in its region" (The Arroyo Group, Planners, Architects, and Associated Disciplines 1981, p. 7). These key urban plans and planning documents had done much to arm the city with a spatially alienating narrative to justify redevelopment decisions.

By the early 1980s, the city had well established a commitment to federal involvement in urban planning and public–private partnerships as a means for spatial alienation. The 1982 City of Santa Ana Annual Report provided a summary of progress in redevelopment dating to 1973, such as producing 275,000 square feet of new retail and office development, over 700 housing units, and rehabilitation projects with an estimated value of over $15 million via the Community Redevelopment Agency's major commercial loan program. It outlined various ongoing projects, such as expansion and redesign of parks, construction of a

Figure 1.7 West Fourth Street in 1983.

Source: Courtesy of the Santa Ana History Room/Santa Ana Public Library.

parking structure, and the Downtown Urban Improvement Project—beautification and public works activities to replace streets and sidewalks (Figure 1.7).

The report also showed that the City Council created four new redevelopment areas designed to eliminate blight through increased public investment, with intentions to fund over $80 million in public improvements over the next decade.

In sum, the DSADP and subsequent key planning documents clearly stated that Downtown Santa Ana, namely Fourth Street, was in the present state in complete disarray, businesswise, physically, racially, ethnically, class-wise, and socially. The assessment of the downtown area was loud and transparent: virtually every single building was dilapidated in some significant fashion, and many businesses were unfashionable, low quality, and too cheap and had a dreadful appearance; consumers, residents, and the overall social scene were criminal, undesirable, and problematic. The DSADP in particular had provided the city with quantitative data on building conditions and anecdotal data on social problems. The city now had plenty of problems to point to so it could "rationally" exercise redevelopment powers and work towards offering opportunities for a utopian society.

Media and Local Representation of Downtown, 1970–1990

Key dominant and community-wide institutions and stakeholders sensationalized the city's prevailing downtown representations of class, ethnic, and racial differences. English-language newspaper media in particular expressed middle- and upper-class anxiety toward the "other," especially Mexican immigrant working classes and the downtown retail spaces that they were beginning to embrace on Fourth Street throughout the 1970s and 1980s.

English-language newspaper headlines would champion proposals for a new facelift of Downtown Santa Ana: "Downtown Plans for Santa Ana to Take off in 1980" (Wong 1979) and "Santa Ana Boasts Ambitious Redevelopment Plans" (Dodson 1984). These and similar newspaper stories pointed out that plans to begin in 1980 included building rehabilitation; razing a long row of bars, hotels, and shops; building demolishment; high-rise office construction and creation of a performance arts complex; addition of parking structures; and the construction of a new mall (Wong 1979, Horovitz 1985).

The English-language newspapers' narrative sensationalized the DSADP's already marginal description of the local community. The routine message of these sources was that the downtown and its people represented a complete slum. This mirrored the city's physical and social blight rhetoric. This perspective was both a response to the growing economic and cultural divisions between social classes and a means to legitimize political and economic activities that fueled the neglected conditions within the downtown.

More direct community redevelopment was essential and long overdue to finally displace the existing community and bring in the middle class. For example, in 1987 a 20-year retrospective on the downtown titled "Santa Ana: A city in transition to become county hub once more," the *Orange County Register* reported, "by the late 1960s, downtown was a commercial slum and home to drunks, drifters and derelicts. The shops where wealthy ladies bought Italian leather shoes were replaced by pawn shops, blood banks and beer bars that catered to illegal immigrants from Mexico" (Snow 1987). The article went on to interview former downtown visitors of the 1950s era to capture memories of the bustling downtown. One interviewee stated that downtown was always happening and "you could hardly find a place to park" (ibid.). The essence of the article's message was that downtown was once very busy, a paradise for the wealthy, overtaken by lesser racial, ethnic, and social classes, and finally moving forward toward reclaiming glory. For the *Orange County Register*, downtown Santa Ana had become, as the City had noted before, "social blight" who replaced the acceptable and commonsense consumers and business types that once again should have claims in the area—White, middle-class, wealthy consumers and the businesses that target them.

The discursive alienation of class, race, and ethnic differences also permeated community-based publications, though more covertly, by ignoring. Representations of the ignoble masses and the spaces they infringed were the antithesis of the social world that the urban elite from middle-class neighborhoods had constructed and were determined to ignore. Local White middle-class community gatekeeper groups of Santa Ana history propagated negative rhetoric of the 1970s and 1980s downtown by not including in their publications images and information of the city's general cultural, racial, ethnic, social, and commercial changes and inundating images of place and people that predated such changes.

Images of America: Santa Ana 1940–2007, which Reed and the Santa Ana Historical Preservation Society (2008) authored, is one such book. While the introduction points out that the city now has cultural diversity and that the city is much more integrated today than in generations past, it clearly obscures the multicultural nature of the city's history. It devotes 127 pages to Santa Ana across five time periods, with 235 images depicting the downtown area, mainly with parades and their participants, commercial/historical buildings and storefronts, streetscapes, county buildings, civic and community leaders, and aerial views of civic buildings. The majority of images, a total of eight, from downtown's oldest section, Fourth Street, date from 1940 to 1949 and 1950 to 1959. The caption for one image notes that 1950s Fourth Street in downtown Santa Ana was "a preferred shopping destination" (ibid., p. 55). From 1960 to 1969 and 1970 to 1979, however, Fourth Street/*La Cuatro* is nowhere to be found, erased from the book. The book's chapter that contains the longest time period, 1980 to 2007, offers one image of a historical building on Fourth Street that once housed the upscale Rankin Dry Goods dating to the early 1900s. The pattern of erasing *La Cuatro*'s cultural character also goes beyond the downtown and Fourth Street, suggesting that this omission may not be accidental. The book also does not depict images of urban streetscapes, cultural celebrations, and daily civic life of the Mexican-American and Mexican immigrant community from the 1970s onward, when this group became a numerically significant Latina/o demographic. It makes no mention of *La Cuatro* as a term and neither *Mexican, Mexican-American*, or *Chicano* appears anywhere in the book. The introductory remarks, "Santa Ana of 2007 is much more culturally diverse than Santa Ana of 1940, but at the same time, the cultures are much more integrated in 2007 than in 1940, as this photographic history of the little town that grew into a big city portrays" (ibid., p. 7), ill suit the content of the book. The book fails to provide any supporting text or imagery to support the claim.

The comprehensive 1994 coffee table book *Santa Ana: An Illustrated History* (1994), makes similar omissions. Written by Diann Marsh, then Board Member of the Santa Ana Historical Preservation Society and "the city's unofficial historian," the book's publication celebrates Santa Ana's 125-year anniversary (Kass 1995). The first chapter is titled "Did Local Indians Live Where Santa Ana is Now Located?" and immediately concludes that they did. The next chapter discusses the local Spanish era, and nine intervening chapters focus on the White history of Santa Ana from 1866 to the 1930s and briefly mention the 1950s and 1960s. The final chapter titled "Renaissance" is reserved for City Council members and local activists that describe the past, the then-present 1990s, and the future of the city, focusing on historic preservation, civic pride, and plans for arts-led redevelopment for the downtown. The book omits the 1970s and 1980s. By omitting these decades, the book effectively erases *La Cuatro* and the working-class immigrants it served. The final chapter has virtually nothing to say about the city's majority Latina/o and growing Mexican immigrant community with the exception of a mention of "Mexican history" and "Hispanic" across two sentences (pp. 160, 165).

For middle-class historic preservationists living in the downtown area that authored or endorsed these key downtown historical publications, the prevailing and selective representations of an ideal downtown overtly avoided mention of fear or disdain of the "other." The discourse of cultural, social, and architectural status quo espoused by the middle classes was not as blatantly marginalizing or sensationalist as those espoused by the DSADP and related documents and dominant newspaper discourse. Instead, these sources simply erase. These authoritative public history circles reorient people's lives and histories and the urban landscapes that they traverse to their own familiar world without acknowledging the world of others.

Latino City Counters the Marginal Discourse

What emerges from the spatially alienated and their allies, collectively defined here as *La Cuatro* business owners, employees, working-class residents, and Spanish-language newspapers, clearly contrasts the narrative of the city's and dominant community's spatial alienation discourse. Fourth Street in the mid-1970s to the early 1980s was at the incipient stages of forming the *La Cuatro's* identity *organically*—a Mexican cultural and ethnic commercial zone that began without cultural-themed intervention, public or private. Mexican business owners, employees, and building owners on Fourth Street began filling lots of vacancies and neglected commercial spaces and as a whole positively valued this new area and era.

Unlike in decades prior to the 1970s, businesses catering mainly to Mexican-Americans and some Mexican immigrants were no longer concentrated on one block on the east part of Fourth Street, away from the contiguous White, luxurious west side (Haas 1995). In the 1880s-1930s, downtown's east area included a pattern of businesses with a Spanish name or an owner of Mexican ancestry, Mexican consumers, and events to celebrate Mexican and Mexican-American identity and culture. By the 1930s, East Fourth Street "was a well-defined Mexican commercial district and it contrasted the more White and higher end West Fourth Street" (ibid., p. 187). In the late 1960s, "no one who worked in downtown spoke Spanish, it was all American"; even the few employees of Mexican ancestry who worked at mainstream stores in the general downtown area typically did not speak Spanish to Spanish-speaking customers because of the stigma attached to speaking the language (personal communication).

By the mid- to late 1970s, this all began to change and changes were an affirmation to Mexican immigrants and Latinas/os generally. *La Cuatro* slowly began forming and increasingly stretching to the west side with a variety of businesses catering almost exclusively to a Mexican immigrant consumer base, namely with specialty businesses, such as affordable clothing stores for men and women, record stores, jewelry stores, entertainment venues such as movie theaters, and a few restaurants. These types of businesses were collectively and slowly reconfiguring a largely bare quarter mile of downtown, slowly filling chronic vacancies and replacing the few remaining mainstream businesses, such as J. C. Penney, on the east side. *La Cuatro* had begun to organically morph from a place for consumption and entertainment of economically well-off Whites to a place where, according to a former councilmember, Latinas/os went on to "save" the downtown by creating "a thriving cultural business district infusing cultural diversity and economically sustaining the downtown" (Marsh 1994, p. 160).

Certainly by the mid 1980s, businesses catering to Mexican immigrants had already filled many vacant stores and sprouted from west to east (personal communication). In 1984, for example, the local *Semanario Azteca* newspaper took notice of this new pattern of businesses and the cultural climate that came with it. The newspaper profiled *La Cuatro*'s vital role for the Mexican community and wrote about the typical businesses there, such as "los cines mexicanos, las tortillerias, los pequeños mercaditos, las discotecas Latinas, restaurantes y cantinas, salones de belleza, zapaterias, estudios fotográficos, imprentas, casas de cambio, etc." ["Mexican movie theaters, stores that make and sell tortillas, Latina/o record stores, restaurants and bars, beauty salons, shoe stores, photography studios, print shops, money exchange stores, etc"] (Velo 1984, p. 3). In 1987, store owners along *La Cuatro* reported that anywhere from 75 percent to 95 percent of their clientele was Latina/o (Dougherty 1987). "And Fourth Street boomed," said Joe Elias, of this time period, a then merchant and current building owner in the area (Cog·Nate Collective 2014)

To illustrate this boom, I describe two key businesses that anchored Fourth Street's unofficial transition to *La Cuatro*. The historic Yost Theater, or *Cine Yost* as referred to by Spanish speakers, owned by a local man, Louie Olivos Sr., from the early 1950s up to the early1980s, and *Calzado Canadá (La Canadá)*, which opened in 1973, are prime examples of entertainment venues and businesses that helped transform the area to *La Cuatro* and draw in large numbers of the Mexican immigrant and Mexican-American population.

The Yost Theater is the oldest theater in Orange County, dating to 1912. Like other theaters in the region and country at the time, for decades this theater racially discriminated against Mexicans and non-Whites by segregating them to the balconies (Bose 2011, Montoya 2002, Zoroya 1990). Under Louie Olivos Sr.'s ownership after 1953, *Cine Yost* put an end to this practice and successfully attracted an audience hungry for Spanish-language live concerts and Spanish-language films, including from Mexico's Golden Age. Mexican immigrant and Mexican-American families, couples, and patrons, irrespective of immigration status and class background, could reasonably afford entrance and proudly sit on the first floor (personal communication). To help ensure a healthy box office, the Olivos family would make rounds on weekends to immigrant farm worker camps and barrios scattered in Santa Ana and Orange County (McLellan 1999). Patrons were also attracted to the theater because it was the only place for Spanish-language live concerts and Spanish-language films in Santa Ana and Orange County. A former patron of the *Cine Yost*, Alfredo Amezcua, recalled, "the Cine Yost … provided many evenings and days of enjoyment to the local Hispanic community. It was … the only show in town" (McLellan 1999).

The *Cine Yost* was much more than simply being the only show in town, however. Louie Olivos Sr. felt it was very important to provide the Mexican community with top quality entertainment. The community benefited from entertainment in what Louie Olivos Jr. referred to as "the crème de la crème" (El Centro Cultural de Mexico 2008). The talent included the most popular Mexican singers and actors of their generation, many of whom maintain icon status today. While the list of genres and musical talents is too long and impressive to list here, the later decades of the Yost included live music concerts from the legendary Vicente Fernandez, Antonio Aguilar, Juan Gabriel, and Las Jilguerillas. These artists commonly offered a widely popular style of traditional music of Mexico, the *ranchera*, which is associated with the mariachi and included themes of love, vengeance, patriotism, rural poverty, and tales related to the Mexican revolution. Vicente Fernandez has two Grammy awards and Juan Gabriel, Antonio Aguilar, and Fernandez have since been immortalized in Hollywood's Walk of Fame.

Unlike the nearby downtown Broadway Theater, which offered only English-language movies, the Yost Theater and the historic West Coast Theater off of Main and Fourth Street, which reopened in the early 1970s, also under Olivos Sr., featured Spanish-language movies and English-language movies with Spanish subtitles. These theaters featured films starring renowned Mexican comedians Cantinflas, La India Maria, Tin Tan, and Andres Garcia, actors whose movies continue to replay today on Spanish-language television stations in the United States.

The *Cine Yost* and West Coast theaters especially drew large crowds to downtown on *estreno de la pelicula* Wednesdays—the debut of a film. Residents who lived in Santa Ana and the region recalled their deep attachment to the theater two decades later, when the Yost Theater reopened for a brief stint for Spanish-language and Mexican cultural programming in 2008. Rodolfo Cordova held back tears and recalled: "It was a beautiful memory for me to be able to return." Teresa Saldivar, who is related to the Olivos family and worked the Yost box office during her childhood, recalled, "I did feel a lot of emotions, it brought back a lot of nice memories when I was little … I basically grew up at the Yost Theater." Lawyer Jess J. Araujo of Santa Ana commented, "Lewis Sr. brought us closer to the big city" (McLellan 1999). The documentary *Santa Ana: La Ciudad de Oro* (El Centro Cultural de Mexico 2008) and Haas (1995) succinctly summarize the significance of the theater for the Mexican community and downtown: *Cine Yost* was the mecca for the Mexican community and made downtown "the happening place" for entertainment from the mid-1950s to the mid-1980s.

Another example of a business that helped create *La Cuatro*'s identity is the family shoe store *La Canadá*, which employed my mother through much of the 1970s and 1980s. In 1973, *La Canadá*—at the time the largest shoe store chain in Mexico—opened a store on the west side corner of *La Cuatro*, historically and now the busiest intersection—Fourth Street and Main Street. As the largest Mexican shoe store in the area, situated in a prime location and surrounded by some vacancies, *La Canadá* was highly recognizable and sought after by many in the Mexican immigrant community. *La Canadá* also attracted lots of consumers from the outset because there were no other shoe stores in the area catering to Mexican families. My aunt Esther Romero Vázquez, a former sales associate, recalls that *La Cuatro*'s identity began to solidify in the area "the minute *La*

Canadá stepped in." My mother, Margarita, was store manager in this period. She remembers *La Canadá* thrived in this downtown environment from the first day of the grand opening, saying "it was crazy!" She recalls,

> *Como llegaba de gente cuando llovia, venia mucha gente de campo para comprar zapato de trabajo, bota de trabajo. Había mucha gente. Cuando era Christmas se esperaba afuera tanta gente, se llenava. Ha veces atendíamos a 10 personas por vendedores, y éramos 6 o 7 venderoes.* [Lots of people would come when it rained, lots of people would come from the agricultural fields where they worked to buy work shoes, work boots. There were lots of people. Lots of people would also wait outside and form a line to get in during Christmas season, it would fill up. Each salesperson would sometimes attend 10 customers at a time and we were about six of seven salespersons per shift.]
>
> [She added in English] We would sell 600 pairs of shoes in one day during Christmas season, which was about $10,000 a day in sales, and we would sell 200 pairs a weekend any other month. Lines of people, from Santa Ana and the general area would wrap around the street corner waiting to purchase shoes.

By the mid-1980s, *La Cuatro* "began being famous," as local business owner Joe Elias put it (Cog·Nate Collective 2014). This positive fame was tied to the Mexican, immigrant, and working-class commercial and cultural identity. Merchants began organizing Mexican cultural festivals and opening additional types of businesses. The city hosted the first annual *Fiestas Patrias*[3] on *La Cuatro* to celebrate Mexican Independence in 1978, but Mexican merchants quickly followed with weekend events such as the first annual 1985 Cinco de Mayo Fiestas. Each celebration was very successful. According to the City of Santa Ana, each routinely attracted anywhere from 150,000 to 200,000 people yearly to *La Cuatro* (City of Santa Ana 2015).

New or more highly visible businesses by this time were also specifically catering to Mexican, immigrant, and working-class families. Business signs and window graphics were typically in one language, mostly in Spanish. The main types of businesses included jewelry stores; travel agencies, almost all specializing in airfare to Mexico and Latin America in general; shoe stores; and general clothing stores. Another business type that became very common was the bridal store, and subsequent hybrid bridal/quinceañera stores (while there were some bridal shops in area in the early 1970s, they catered to Whites). One such store was as *Minas Bridal*, named after the Mexican business owner. Bridal shops were popular for specializing in merchandise for multiple religious ceremonies, such as baptisms, first communions, and toward the late 1980s to the present, more widely recognized for selling quinceañera merchandise and related event planning. One bridal shop owner recalled that bridal business began to flourish towards the end of the 1980s:

> People who worked [in this industry]—I was one of them; I came to help my sisters—saw that it was a business, and that there was work for everyone,

> and that there was a lot of people, and demand for this kind of industry. So [we each said] let's open another store. May family had one store, and then we opened another, and another, and another. And other workers, when they knew how more or less the business went, they opened their own bridal shop. They invested and they opened their own bridal store. So of the people that are here now, that I know, almost all of them were ex-employees of bridals, and now they have their own store—some have one store, others have two … I counted all of them on Fourth Street one day, something like 30, 35 [bridal stores]. (Cog·Nate Collective 2014).

The social environment was also active. For instance, record stores, which were also quite common in the area, routinely played loud *ranchera* and Spanish pop music, blasting through fifteen-inch speakers placed outside near entry doors. Some stores hired clowns on weekends and during sales events to attract customers, offering children balloons to entice families to shop. Cars filled the narrow streets, sometimes with loud Spanish music emanating from their speakers, and parking was very hard to come by. Sidewalks were filled with families and often mothers walking with a stroller and children in tow. "I remember that when I didn't bring food I had to go out for lunch, and I had to think about it, because you had to walk sideways, and it took a long time [to arrive at my destination to purchase food]" because the sidewalks and area was very crowded, reflected one area bridal business owner (Cog·Nate Collective 2014).

Immigrants would continue to carve out their contributions to the thriving downtown in the midst of the harsh realities of criminalization that threatened this social and economic vibrancy. The random sweeps of the area by the federal Immigration and Naturalization Services (INS, now known as Immigration Customs Enforcement, ICE) from at least 1976 to the mid-1980s was a reminder to immigrants specifically and the Latina/o community in general of the city's anti-immigrant expressions about them and role in slowing sales on *La Cuatro*. Merchants understood that minimal foot traffic and slow sales are key indicators of a lagging commercial economy. While the local chapter of the League of United Latin American Citizens and the Santiago Club, a local Latina/o service organization, publically praised Police Chief Raymond C. Davis[4] in 1985 for refusing to assist the INS in sweeps (Rose 1986), the INS sweeps of the area under the city's watch had already created trauma for families and the local businesses.

Yet, business employees commonly did what they could to buffer against the INS and the impact that they could have on families and the local economy. For instance, *La Canadá* was one of the businesses in the area that would routinely lock the store doors to keep the INS from entering when they were around, thereby shielding Mexican customers from officials, if only for a moment. Business employees looking out their storefronts or doors would sometimes catch glimpses of a group of INS officials, often wearing green uniforms and sunglasses, walking in teams up and down *La Cuatro*'s sidewalks. According to my mother, these scenarios, some of which I witnessed firsthand, would set off urgency among many business employees to protect fearful families from potential eminent harm to families. "*Sierren las puertas. No se preocupen*," [Lock the doors! Don't worry!] her employees would instruct each as they skirted past frightened families trying on shoes to lock the front doors. "*No se asusten, no pueden entrar*" [Don't be scared, they don't have legal rights to enter]. The shoe store employees, all young adult Mexican immigrants who had immigrated as children, empathized with their clients. They believed the INS was inhumane and had no legal right entering the store to profile families, and this compelled them to protect their customers (personal communication). These sweeps were indeed traumatic for families and employees alike.

The INS sweeps and the fear of potential sweeps had some consequences on the local economy. While it is unclear how many arrests the INS made and to what extent patrolling impacted sales during the 1970s and 1980s, Carlos Ferrusca, owner of then Ferrusca Photo Studio and Bridal Shop on *La Cuatro,* offers some perspective to these questions. He told the *L.A. Times* that INS profiling and arrests of Latinas/os in the fall of 1986 contributed to the 60 percent decline in sales at his store (Christensen 1987). Patrolling continued months later. In February 1987, the *L.A. Times* reported that federal agents arrested 85 people suspected of being "illegal" immigrants in the downtown (ibid.). Merchants were generally convinced that INS raids were a sign of the city's power to ignore criminalization and indirectly facilitate displacement of Mexican consumers to other cities due to fear of arrests, all helping slow the downtown economy.

Notes

1 The DSADP offered recommendations along three categories (land use, consolidation of blocks into larger blocks, street realignments and closures) and due to space, not all these are covered in similar depth here.
2 Evaluation categories were, from best to worst, 1 = Standard, 2 = Deteriorating, and 3 = Substandard.
3 Haas (1995) shows that a Mexican Independence celebration was organized by Mexican organizations in the downtown area before this period, however, in Birch Park in 1920.
4 According to a March 29, 1986 *Los Angeles Times* story, Davis recently "took over a newly created post of deputy city manager in charge of the Police and Fire department."

References

Adams, V. Ruth. 1963. *Preliminary survey of the City of Santa Ana*, edited by Santa Ana Public Library. Santa Ana, CA.

Arellano, Gustavo. 2008. *Orange County: a personal history*. 1st Scribner hardcover ed. New York: Scribner.

Bose, Lilledeshan. 2011. "The Yost Theater is ready for its closeup." *Orange County Weekly*. www.ocweekly.com/2011-07-28/music/the-yost-theater-santa-ana-olivos-chase-family-dennis-lluy/2/.

Bricken, Gordon. 2002. *The Civil War Legacy in Santa Ana*, edited by Santa Ana Historical Preservation Society, Living History Books Project. Santa Ana, CA.

Canalis, John, Bradley Zint, and Michael Miller. 2015. "Henry Segerstrom, South Coast Plaza and arts center pioneer, dies at 91." *Daily Pilot*, February 20, 2015. www.dailypilot.com/news/tn-dpt-me-henry-segerstrom-south-coast-plaza-pioneer-dies-at-91-20150220,0,4006018.story.

Cannon, Lou. 1978. "Conservatism, record of integrity fade in Orange County." *The Washington Post*, February 12, 1978.

Christensen, Kim. 1987. "85 arrested in Santa Ana immigration raid; downtown merchants upset." *Orange County Register*, February 12, 1987.

City of Santa Ana. 1982. Community Development Application: Fiscal Years 1979–1982.

City of Santa Ana. 2009. City of Santa Ana General Plan: Housing Element 2006–2014.

City of Santa Ana. 2015. "Cinco De Mayo Festival, City of Santa Ana." Accessed June 20. www.ci.santa-ana.ca.us/parks/cinco/.

Cog·Nate Collective. 2014. Borderblaster (SNA): La Cuatro (I-IV). Santa Ana: Cog·Nate collective.

Community Redevelopment Agency. 1973. *Redevelopment Plan City of Santa Ana Redevelopment Project Area*, Santa Ana, CA.

Dermengian, M. Sam. 1955. *A market analysis and trade study of Santa Ana, California*. Santa Ana, CA: Santa Ana Chamber of Commerce.

Diaz, David R. 2005. *Barrio urbanism: Chicanos, planning, and American cities*. New York: Routledge.

Dodson, Marcida. 1984. "Santa Ana boasts ambitious redevelopment plans." *Los Angeles Times*, September 24, 1984, 3–5.

Dougherty, Kathleen. 1987. "Fourth Street—A gem of an area—shops concentrate downtown to sell all facts of jewelry." *Orange County Register*, March 31, 1987.

El Centro Cultural de Mexico. 2008. El Cine Yost en La Ciudad de Oro. Santa Ana: El Centro Cultural de Mexico.

Fashion News. 2015. "South Coast Plaza predicts $1.5B in sales | SGN Group." http://sgn-group.com/fashion-news/south-coast-plaza-predicts-1-5b-in-sales. Accessed October 20, 2015.

Garcia, Mary. 2007. *Santa Ana's Logan Barrio: its history, stories, and families*. Santa Ana, CA: Santa Ana Historical Preservation Society.

Goddard, W. A. 1969. "Fabulous fifties." In *Santa Ana's 100 years: Prelude to Progress, 1869–1969*, edited by W. A. Goddard and Francelia S. Goddard, 92–100. Santa Ana, CA: Goddard, W. A.

Haas, Lisbeth. 1981. *The bracero in Orange County,California: a work force for economic transition*. La Jolla, CA: University of California San Diego. Texte imprimé.

Haas, Lisbeth. 1995. *Conquests and historical identities in California, 1769–1936*. Berkeley, CA: University of California Press. Texte imprimé.

Harwood, Stacy, and Dowell Myers. 2002. "The dynamics of immigration and local governance in Santa Ana: neighborhood activism, overcrowding, and land-use policy." *Policy Studies Journal* 30 (1):70.

Horovitz, Bruce. 1985. "New name, new face for Fashion Square." *Los Angeles Times*, May 22, 1985. http://articles.latimes.com/1985-05-22/business/fi-16915_1_fashion-square.

Irazábal, Clara, and Ramzi Farhat. 2008. "Latino communities in the United States: place-making in the pre-World War II, postwar, and contemporary city." *Journal of Planning Literature* 22 (3):207–28. doi: 10.1177/0885412207310146.

Kass, Jeff. 1995. "At home with history: publishing: a Santa Ana resident has written a book about the city's architectural and civic past. She not only writes about it, she lives with it in the French Park district." *Los Angeles Times*, June 1, 1995. http://articles.latimes.com/1995-06-01/local/me-8313_1_santa-ana-history-room. Accessed April 12, 2016.

Kennard, Delahousie and Gault. 1974. *Downtown Santa Ana Development Plan*.

Lefebvre, H. (1991). The Production of Space. Cambridge, MA: Blackwell. First published as La Production de L'espace (1974).

Marsh, Diann. 1994. *Santa Ana: an illustrated history*. 2nd ed. United States of America: Heritage Publishing Company.

McLellan, Dennis. 1999. "The show must go on." *Los Angeles Times*, January 3, 1999. http://articles.latimes.com/print/1999/jan/03/news/cl-59846. Accessed April 12, 2016.

Mead, Julia C. 2003. “Memories of segregation in Levittown.” *The New York Times*, May 11, 2003.
Montoya, Margaret E. 2002. “A brief history of Chicana/o school segregation: one rationale for affirmative action.” *La Raza Law Journal* 12:159–71.
Orange County Historical Society. 2015. “A brief history of Orange County California.” www.orangecountyhistory.org/history-brief.html. Accessed June 13, 2015.
Reed, Roberta A., and Santa Ana Historical Preservation Society. 2008. *Images of America: Santa Ana 1940–2007*. Charleston, SC: Arcadia Publishing.
Rose, Andy. 1986. “Santa Ana: Davis named as 1985 Police Chief of the Year.” *Los Angeles Times*, March 29, 1986. http://articles.latimes.com/1986-03-29/local/me-1338_1_police-department. Accessed June 12, 2015.
Ruth, Krushkhov, D. Arthur Little, and J. D. Faustman. 1965. City Center Studies and Plan: Technical Report No. 5. In *General Plan Program: Santa Ana, California*.
Santa Ana Planning Commission. 1965. *Proposed General Plan for Santa Ana*.
Smith, N. (1996) The New Urban Frontier: Gentrification and the Revanchist City. London: Routledge.
Snow, Anita. 1987. “Santa Ana: a city in transition to become county hub once more.” *Orange County Register*, June 7, 1987.
The Arroyo Group, Planners, Architects and Associated Disciplines. 1981. City of Santa Ana General Plan Program Alternatives Evaluation.
U.S. Census Bureau. 1980. Census of population and housing. PHC80-2-67, edited by U.S. Department of Commerce: Washington, D.C.
Urban Futures, Inc., and Planning Department City of Santa Ana. 1975. Santa Ana Housing Plan: Background Report.
Velo, Fernando. 1984. “El pez mas grande se come al chico.” *Semanario Azteca* 4 (July 1).
Warner, Kee. 2012. “Placing barrios in housing policy.” In *Latino urbanism: the politics of planning, policy, and redevelopment*, edited by David R. Diaz and Rodolfo D. Torres. New York: New York University Press.
Wong, Herman. 1979. “Downtown plans for Santa Ana to take off in 1980.” *Los Angeles Times*, December 22, 1979.
Zoroya, Gregg. 1990. “Hispanics: immigrants bore brunt of discrimination.” *Orange County Register*, August 26, 1990.

2 The Politics of Redevelopment and Resistance to Eminent Domain, 1980s

In the early and mid-1980s, Santa Ana's government promoted a more democratic direction in community development than had characterized efforts in the 1970s. The city began to conduct community outreach activities to listen to and understand the community. At the same time, the institutionalization of disenfranchisement of the growing Latina/o community advanced, particularly through plans for eminent domain, and especially in the downtown area. The city's actions outraged activists and business and property owners. Downtown businesses began to act against the redevelopment climate. This chapter explores how the city reached out to the community and community reactions to these efforts. Focusing two high profiled cases at the time, I ask: to what extent do struggles against eminent domain address a broader grassroots climate advocating for city reinvestment for the Latina/o community?

The City Promotes Democracy and Personal Responsibility

In the summer of 1981, the city held a historic meeting in what was believed to be "*[la] primera vez que un Alcalde paraticipa abiertamente con sectores minoritarios*" [the first time a Mayor openly participated with the minority community] (*Semanario Azteca* 1981a, p. 1). Mayor Gordon Bricken held a roundtable breakfast meeting to learn firsthand what Latina/o leaders were thinking and concerned about. In attendance were about 30 Latina/o leaders and professionals from numerous civil rights, labor, and business organizations and other groups. Groups in attendance that were representing the Latina/o community included the League of United Latin American Citizens (LULAC), Latin-American Chamber of Commerce, and Laborers Local 652 (ibid.). In another community meeting at the start of the following year, City Manager Arnold J. Wilson held a press conference stressing that he wanted to establish communication with the community and make redevelopment more responsive, a reoccurring theme from the previous meeting with Latina/o leaders (*Semanario Azteca* 1982a). He emphasized that the city was receptive to work on contentious issues, such as park upgrades. Another example of the city's democratic practice was a community meeting with over 200 people to educate the public about the upcoming 1982 General Plan and to solicit feedback.

While the city was promoting democracy and listening to the community, city officials were also becoming uneasy with the idea of addressing redevelopment that embraced equity and the poor. During a May 1984 City Council meeting, the Council voted on a proposal for $100,000 in federal funds for no-interest loans for families that would be displaced from a redevelopment project (*Semanario Azteca* 1984c). The measure narrowly passed four to three and is indicative of mixed views on resource allocation for the poor and displaced. For example, City Manager Robert C. Bobb's final attempt to sway the Council in a "yes" vote pointed out that the Council would be sending the community the wrong message if the policy did not pass: "*si la medida no es aprobada, el concejo estara enviando un mensaje muy sensitivio a la comunidad. La cuidad debe asumir la responsablidad de proteger a nuestros residents*" [City Council will be sending the community the wrong message if it does pass the measure. The city should be responsible in protecting our residents" (*Semanario Azteca* 1984c). Patricia McGuigan, who had publicly voiced months earlier that she did not believe it was the city's responsibility to assist displaced families, ended up voting in favor of the measure. Former Mayor Bricken was concerned that some residents would be upset that federal city dollars would be allocated for displaced residents, some of whom would be unauthorized. Acosta, the city's first Latina/o councilmember, voted "no" on the measure, saying that he was "firm" in his opposition against the measure to help the community and that the city should not waste money on poor families (ibid.).

Despite these more inclusive planning practices and redistributive policies, the city continued to institutionalize its dominant discourse for the downtown. For instance, in 1984, Deputy City Manager Rex Swanson told the *Los Angeles Times*, "Give us three more years. We're going to establish a real downtown in Orange County" (Dodson 1984). Swanson suggested that businesses catering to Latinas/os and Mexican immigrant consumers must vacate the area. He called them a barrier to a "real" downtown. Latina/o businesses were "marginal" and had been able to operate downtown largely because of currently depressed rents, which were bound to jump when buildings were improved and renovated, he added. The three-year redevelopment process would work to subtly phase out some of the businesses that were too particular and unwelcoming and replace them with those that catered to suburban patrons, allegedly creating a more inviting, safe, and inclusive mixed area. "In my opinion all suburban shopping centers are dull and all the Hispanic shopping centers are exclusionary. We want to mix them and make people comfortable. There's a thin line between urban flair and urban fear," he said (ibid.).

Other policy-makers expressed that, while new plans for downtown businesses were necessary because Mexicans who shopped there were too poor, achieving this vision posed serious challenges if they were to work and serve the present community. For example, council member Gordon Bricken, speaking on behalf of the city, declared these sentiments in a 1984 interview with a member of the city's historic preservation society:

> Economically, [Santa Ana's downtown] has not been successful. The government center has simply not materialized the numbers we had anticipated. We had a consultant hired by CD [Community Development] to determine what the market was for downtown, and he came back and said that the market was all Mexican and had a median income of $13,000. I didn't need a $30,000 consultant to tell me that. So what does it mean? Do we try to bring people here who earn more money or hang to the people here now? (Londoño and González, 2015)

For David Ream, Santa Ana's executive director of economic development in 1984, the future looked much brighter than the 1970s and the city needed to take swift action to boost property values and keep pace with suburban neighbors. In 1984, he recalled that downtown property values and tax revenue were very low during the 1970s and was hopeful that eager developers would ultimately help turn this around. "It was disastrous back then [1970s]. This [downtown] was the only place where the assessor's office was reducing property values and decreasing taxes, and this was during Orange County's property boom. That's bad" (Dodson, 1984, 4). Showing optimism, he added that developers in the 1980s were instead "standing in line" to build in downtown Santa Ana (ibid.). Redevelopment meant maximizing property values and taxes and developers were more than ready to help the city in this respect, Ream added.

Eminent Domain and the Equity Question

The city banked on eminent domain to maximize the tax base while giving much less attention to distribution and equity than activists were calling for. The power of eminent domain would also remind local merchants that the city was prepared to displace businesses catering to existing communities and appeal to new kinds of socioeconomic communities.

The Santa Ana Community Redevelopment Agency, whose members also sat on the City Council, sustained their faith in eminent domain and the private market in the mid-1980s and continued to condemn private property to build public projects. "I can't imagine that redevelopment would work without the power of eminent domain," said Rex Swanson (Snow 1987a). The power of eminent domain would allow the agency to buy small parcels in the downtown, package them into larger parcels, and turn them over to developers for construction of blocks of commercial projects that would improve the city's tax base, he added.

City officials maintained that economics, not racism, drove redevelopment decisions. In a June 1984 City Council meeting to vote on the city's decision to sell property on First Street to developers to make way for a shopping center, City Manager Bobb commented, "Tonight, we are making a business decision. Emotions don't count and we also can't make this a racial issue. We would be showing a negative attitude against the business sector if we don't approve contracts with developers" (*Semanario Azteca* 1984a). Councilmember Bricken could not hold back, mentioning a racist undercurrent that prevailed in the general community. He said, "The future of the city has been in constant discussion and this is no secret. It is not about Latinizing downtown Santa Ana, just as it is not completely true that Anglos would like to get rid of Mexicans" (ibid., p. 4). Activists understood this racist undercurrent and began to mount protest.

Residents and Merchants Protest

Many activists expressed their outrage with the city's dominant views toward the Latina/o community and redevelopment and participation discourse and practices. In 1977, seven newly formed neighborhood organizations, representing mainly working-class Mexican-American barrios and some middle-class White neighborhoods, came together with Civic Center Barrio and formed the Santa Ana Neighborhood Organization (SANO) to oppose and question redevelopment activity stemming largely from the DSADP. Community organizers from Oakland spearheaded SANO's formation as an Alynksy-style organization (personal communication). Activists pressed that the "*el plan [general] estaba desenado de una forma en que iba a destruir una comunidad vecinal completa*" [the [general] plan was designed in a manner to completely destroy neighborhoods] (*Semanario Azteca* 1982c, p. 11). Destruction largely meant that the city was using or planning to use eminent domain to physically demolish and replace existing housing and commercial options on or near Fourth Street. Others understood that redevelopment was a way for the city to institutionalize racism and rid the area of minorities in favor of boosting image and private interests. "*Estamos viviendo una politica racista, especialente contra los Latinos. Ellos-los oficiales de Santa Ana—quieren traer grandes corporaciones, quieren vender una excelente imagen de esta ciudad y para ello estan discriminando*." [We are living among racist politics, especially against Latinas/os; Santa Ana city officials are discriminating because they wish to attract large corporations and promote an excellent city image], added one resident (*Semanario Azteca* 1984b).

Activists attended City Council meetings and city-sponsored community meetings, and protested in front of redevelopment sites. As a whole, though, activists were in favor of redevelopment so long as the city provided ethnic minorities and the poor with reasonable housing options (*Semanario Azteca* 1982b). Sam Romero, President of SANO, expressed some of these views to city officials during a community meeting:

> *Nosotros no nos oponemos a que exista un proceso de remodelacion, lo que si estamos en contra es de que no existan facilidades para adquirir viviendas a precios moderados y que sean dignas de habitar. Este es un proceso en el cual los mas perjudicados seran los residents de las minorias ya que se quiere implementar areas comerciales y vias de transito rapido los cuales sin duda afectaran a nuestra gente*. [We are not against redevelopment. We are against the lack of options to obtain quality and affordable housing. Ethnic minorities will be the most affected by commercial and transit redevelopment projects which would all undoubtedly affect our people.] (*Semanario Azteca* 1982c, p. 10)

Activists were also infuriated with the city's community participation practices, which they felt were smokescreens for democracy and lack of proper attention to basic neighborhood infrastructure. For example, more than 300 attended a City Council meeting and protested the General Plan and the tactics and activities that the city used to suppress resident feedback and involvement. *El Quetzal* (1977) newspaper noted that some city officials believed this highly attended meeting "to be the largest 'grass roots' crowd in memory" (p. 1).

Local community organizations were part of the protest and vocalized numerous concerns, such that the plan was created without "real community participation and human dignity," citing as an example a recent general City Council meeting that would have allocated residents time to speak up on the plan but was rescheduled to an earlier time slot to discuss plans without the community in attendance (El Quetzal 1977). The community continued to be vocal with the city's community participation activities into the 1980s. Some activists were now expressing that city officials were often deliberately vague discussing specifics during their community meetings about the General Plan's goals and objectives would benefit or keep stable the Latina/o community in the areas of housing and commercial activity, among other topics (*Semanario Azteca* 1982c).

Other activists were upset with political representation; they accused city officials of socially constructing urban decay. In a community meeting, for instance, a woman in attendance asked city staff how long they had been living in Santa Ana, suspecting that the staff was out of touch with locals, and one staff member responded, "six months" (*Semanario Azteca* 1982c). In reaction, some in the crowd laughed, asking how anyone can carry out city plans without having much knowledge about the city in the first place. Activist Rosa Maria Prado sounded off to officials that they were inaccurately depicting her city as deteriorating. "*Nuestra ciudad el la cual hemos vivido por muchas decadas, por varias generaciones no es una urbe decadente*" [Our city, which we have lived in for many decades and across many generations is not deteriorated] (*Semanario Azteca* 1982b, p. 7).

Activists affirmed they would take a proactive grassroots stance against the city. Some residents would not relent fighting for the local community or vacate the downtown, proclaiming that *"Las intenciones son botar a los Mexicanos pobres de Santa Ana, y eso debemos impedirlo, cueste lo que cueste.*" [The intentions are to get rid of poor Mexicans from Santa Ana, and we should prevent that at any price] (*Semanario Azteca* 1984b). Another resident argued, "*No vamos a irnos, que lo hagan los incapaces, los que quieren hacer una ciudad para los ricos, par las corporaciones.*" [We are not leaving, let the inept leave, those who want to make a city for the rich, for corporations] (Semanario Azteca 1982b, p. 7).

SANO achieved numerous victories before it disbanded by the mid-1980s. One such significant victory was a 1978 $2.6 million community budget prepared by SANO. The budget included redevelopment items and dollar amounts that were missing in the city's original budget, such as specific and higher dollar amounts for street and neighborhood improvements in each community development target area. SANO's grassroots activism and critical analysis of the city budget ultimately pressured the city to rethink city politics. SANO had successfully altered some budget decisions to benefit neighborhood priorities.

La Cuatro merchants also had similar outrage during the early to mid-1980s. For example, merchants were identifying with the broader Latina/o resident and activist community and feeling threatened by redevelopment they believed would displace and gentrify the area. For example, some merchants on *La Cuatro* were convinced that the true intentions of the early 1980s General Plan were to wipe out their businesses and Latina/o shoppers from the downtown (*Semanario Azteca* 1981b). The article *Amargura en Santa Ana* [Bitterness in Santa Ana] (*Semanario Azteca* 1984a) reported in 1984 that *La Cuatro* merchants also felt undervalued and humiliated and public enemy number one to be displaced through redevelopment activity. One merchant expressed: "*Nosotros estamos en el centro de la ciudad por varios anos, cuando ningun anglo queria venir aca, y ahora que se esta remodelando el sector, se nos trata de quitar de nuestas tiendas.*" [We have been in the downtown for various years and not one Anglo wanted to come down here, and now there are efforts to take our stores away during revitalization] (ibid., p. 4). The article further reported that merchants repeatedly were arguing, "*El 65% de la poblacion de Santa Ana es latina no es justo que la ciudad nos desprecie, nos humille en forma arrogante y nos desloje*" [Sixty-five percent of Santa Ana residents are Latina/o, and it is not right that the city devalues and embarrasses us by portraying us as ignorant and displace us] (ibid., p. 4). Merchants were also reporting that the city had avoided communicating community development plans to them in order to avoid their resistance.

The Pulido Family Resists Eminent Domain

Arguably the highest-profile business battle and victory in Santa Ana eminent domain redevelopment history is the case of the Pulido family, who avoided losing its muffler shop, Ace Muffler Shop, to eminent domain when the city was planning for a shopping center (*Semanario Azteca* 1985). Miguel Pulido, who in the mid-1980s was a young Mexican immigrant and manager for his family-owned ACE Muffler Shop and would become a city councilmember in 1986 and mayor of Santa Ana in 1994, became embroiled in fighting against the city to keep his father's business in the proposed shopping center. The fight was important to the family because eminent domain would threaten the family's economic livelihood. Pulido and his family had the support of the business community and other grassroots organizations, many Latina/o-led, to apply the political pressure necessary to stop city plans to condemn his family's auto shop (Snow 1987b). For many Latina/o leaders, the fight symbolized defending merchants and businesses against what they believed was racial discrimination to "rid downtown of Hispanic businesses" (personal communication).

The Pulido family campaign was multidimensional, taking on the form of directly securing personal economic interests and indirectly and more distally promoting economic interests for people whose business or properties could likewise be taken in eminent domain. The campaign also promoted civic participation in electoral politics, which Pulido argued was the best way for future generations to avoid eminent domain. Pulido stated:

> *Vamos a luchar para quedarnos, registraremos nuevos votantes, propondremos concejales en la proxima eleccion y nos organizaremos para que jamas vuelvan a dar 'atole con el dedo'. Solo el voto nos dara el poder necesario para vencer y por ello tenemos forzosamente que unirons pues si hoy dia el afectado fue mi negocio o mi casa? Quien dice que mana no sera el suyo?* [We will fight to remain, we will register new voters, we will nominate councilmembers in the upcoming elections and we will organize to avoid being deceived.... Voting is the only way we will obtain the power to overcome and we have to unite because, while today it may be my business or house that are affected, who says that you will not be affected tomorrow?] (*Semanario Azteca* 1984a, p. 4).

The mounting support for the Pulido family from the Latina/o community prompted the city to hold a press conference with Miguel Pulido and a number of high-ranking city officials. The press conference was the city's way to explain to the general public that the intentions to use eminent domain were not personal against the Pulidos and was not about racial discrimination. "There are thousands of merchants who would like a similar opportunity," Mayor Dan Griset commented (*Semanario Azteca* 1985, p. 1). Councilmember Robert Luxembbourger assured the public that the case was not "anti-Hispanic" (ibid., p. 1).

In the spring of 1985, the city agreed to let the Pulido family business remain after years of community pressure (Snow 1987b). The Pulido family became the first Mexican immigrant business owners in the downtown area to organize successfully against eminent domain. This highly publicized fight was another reminder to the city that activism against ongoing redevelopment generally and eminent domain specifically was likely not going away anytime soon.

The Pulido fight was about the family's economic interests and divorced from community equity principles that the general community had been advocating for. Pulido tailored his family's fight narrative against the city as an issue that the entire community needed to empathize with because any one business or property could also fall prey to eminent domain and risk personal loss. While he likewise appealed to the general community to organize and get involved in electoral politics, securing policies and resources for the downtown area to support existing business were not part of his agenda.

The Pulido campaign narrative could be better understood with the slogan "the business or property that you save will be your own." This emphasizes the need to maximize or secure personal gains, a purpose associated with liberal thought. In his case this is private capital. This personal economic interest contrasts the community equity narrative (Lejano and Wessells 2006). We can see the contrast by considering the following. A multi-level narrative combining personal and community wide intentions might be: "the businesses we save together might include your own." The underlying perspective speaks to the need to distribute benefits that are inclusive of others and not only personal. This narrative can combine financial benefits and community-wide aspirations and needs.

The Case of Fiesta Marketplace

The highly publicized Pulido battle was underway when a group of business and property owners partnered in response to the city's designation of the eastern edge of *La Cuatro* for eminent domain and blocks of commercial redevelopment. Initially unopposed, the city decided that the best land use for the area would be a "festival shopping" center for which it hoped to begin construction in 1985. This type of festival-themed redevelopment was popular nationwide to redevelop cities with waterfronts, and Baltimore and Seattle had recently completed projects of this type (Schwartz 1989). The city began to aggressively send out about 200 requests for proposals to developers around the country inviting them to submit plans.

While the city was moving steadily forward, a local Mexican-American businessman on the proposed site for the festival shopping center would work on a deal with the city to both protect his economic interests and distribute some benefits among and for the local community. Robert Escalante, a recently minted businessman from Santa Ana who had just purchased commercial property to take over Packard Restoration, a high-end classic car restoration business, had received a letter from the city telling him that his property was subject to eminent domain to make way for the center (personal communication). He needed to think and act quickly. He attended community meetings held by the city that discussed eminent domain plans and the festival shopping center and held informal discussions with key city leaders, such as the city councilmembers and executive director of economic development, pleading with them to change their minds (*Semanario Azteca* 1984a). Escalante advocated to keep his property and informally proposed to develop an ethnic-themed festival public–private open mall shopping center on the growing commercial success of *La Cuatro*. San Antonio's Riverwalk and the Los Angeles Plaza Historic District were examples that Escalante's idea could resemble.

The city did not seize Escalante's property and worked with him to plan the ethnic-themed shopping center. The city acted swiftly. Mayor Daniel Griset championed Escalante's vision in city hall and halted the process of recruiting a developer for the original project idea (Schwartz, 1989; personal communication). Escalante maintains to this day that the mayor was his steadfast supporter. While the mayor's personal motives are unclear to Escalante, he believes that the timing of Pulido case helped his interests tremendously. Escalante maintains today that the city approved the Fiesta Marketplace in large part in an effort to deflect any additional accusations of racial discriminatory redevelopment (personal communication).

City officials could have suspected, however, and quite reasonably, that there could be political or legal backlash from developers if the city awarded a major redevelopment project to Escalante and not an actual developer. To gain some legitimacy as a developer and at the encouragement of some city officials, Escalante formed The Greater Eastern Development Corporation and won the redevelopment bid (City of Santa Ana 1985a).

Escalante immediately partnered with others who were also facing eminent domain in the original project area. Local business owner and friend José Ceballos, a Mexican immigrant who owned property and two record stores specializing in Spanish-language music and books, was the first person that Escalante recruited. Sometimes meeting in bars the city wanted to raze, the friends crafted and proposed a provisional plan to the city which they would be responsible for executing and managing.

By Escalante's and Ceballo's admission, the "developer" role that they both were taking on was beyond their scope and pockets, and they needed to broaden their partnership with members who could give them a better chance at succeeding. Escalante was quite young, in his early thirties, a novice businessman, and by no stretch a developer, he said. Ceballos, also in his early thirties, was still learning how to run his stores, the first of which he opened on *La Cuatro* in 1977. Escalante and Ceballos understood quite well that the work ahead would require capital and commercial development expertise that the men did not have. Escalante and Ceballos collaborated with City Manager Robert C. Bobb and took the first step to discuss ways to satisfy the city's initial financing requirements, which included a $50,000 Good Faith Deposit (Community Redevelopment Agency of the City of Santa Ana 1985). Escalante subsequently invited seasoned local real estate investor Allan Fainbarg, who also had property in the proposed site, to join the partnership and satisfy the city's financing requirements. Fainbarg immediately recruited his son-in law Irving Chase, also a commercial real estate investor. Fainbarg's and Chase's capital and general commercial real estate background seemed to the partners to be suitably sufficient to strengthen the partnership. Fainbarg and Chase did not have festival shopping or ethnic festival shopping redevelopment experience, but this mattered little to Escalante, Ceballos, and the city. Politicians agreed to the Fiesta Marketplace proposal after Fainbarg and Chase committed financing. It was as if a golden real estate and financial opportunity landed on Fainbarg and Chase without much political effort from the start. As a case in point, Chase admitted in a newspaper article that he was surprised when Bobb and Ream agreed to the project and told him to put together a more sophisticated proposal than what Escalante and Ceballos had originally drafted (Schwartz 1989).

Almost immediately, Fainbarg and Chase gained command of the project. On behalf of the partnership, Chase hired a consultant whom he was acquainted with to draft numerous planning and legal documents related to the project, such as project plans, a HUD grant, memorandums of understanding, and related documents (personal communication). The plan called for 50,000 square feet of new retail/office space and 50,000 square feet of rehabilitated retail/office space across four blocks. A new multiplex movie theatre with adjoining retail shops and a food court and a two-level building with ground floor retail and upper floor office space would form part of the new construction. The plan was also to renovate the Yost Theatre for entertainment purposes. The development was supposed to incorporate high-quality site planning and architectural design and a pleasing, safe, and well-maintained environment (Community Redevelopment Agency of the City of Santa Ana 1985).

In the summer of 1985, the economic and decision-making power swayed toward Fainbarg and Chase. The city and Fiesta Marketplace Partners signed a disposition and development agreement. Fainbarg became the managing general partner, Chase the general partner, and Escalante and Ceballos limited partners. Rangel, who had been part of the original informal discussions, would formally join the partnership as a limited partner within months. In 1989, the estimated $10 million Fiesta Marketplace opened.

The Packaging of Fiesta Marketplace

The partners packaged Fiesta Marketplace with both private entrepreneurship interests and intentions to address the city's Latina/o community. The partners reasoned in the Fiesta Marketplace grant and related documents that the project would allow the partners to stay in business and they could only develop Fiesta Marketplace with heavy city and federal assistance; the city and federal government complied quite generously. One month after receiving the partners' proposal for the marketplace, in the summer of 1985, the city issued and sold $5.8 million in tax-exempt bonds to the partners for the acquisition, construction, and rehabilitation of Fiesta Marketplace and related facilities, funds that had been authorized for Escalante's Greater Eastern Development Corporation six months prior (City of Santa Ana 1985a). The partners also gained financially from the federal government. The city submitted a federal Housing and Urban Development (HUD) Urban Development Action Grant (UDAG) application on behalf of the Partnership and secured $680,000. According to the draft grant application,[1] the grant would be the final financing needed to complete the project. The funds would offer suitable financial incentives to prospective developers who may wish to acquire and upgrade the commercial facilities within the area. The application demonstrated need by arguing that buildings in the project area needed to be retrofitted to be earthquake-safe and in compliance with state and city law. Yet, the eastern edge of *La Cuatro* had a shortage of reasonable tax gains and partners had limited financial capacities to comply, the application added.

The partnership also argued that Fiesta Marketplace combined personal economic development benefits with the intention to structure options for economic benefits for the general business community who wished to join them. For instance, for Escalante and Ceballos, Fiesta Marketplace represented an opportunity to gain financial security for their family and pay for their children's college education "when they grow up." More broadly, project planning–related documents noted that the real estate arrangement would give the partnership and general property owners in the project area the option of holding onto their buildings and buying into the partnership. Those who would choose to leave would have to sell their buildings to the city through the city's powers of eminent domain, and in turn the city would sell the buildings below market value to the partnership and maintain an equity position in the project, with the partnership retaining a percentage as a loan from the city (City of Santa Ana 1985c). Another element of these dual intentions reflected trickled-down benefits for the broader Santa Ana Mexican community, which was described as the "massive influx of both legal and illegal immigrants from Central and Latin America" (ibid., p. 4). These benefits were providing jobs and helping address the city's high unemployment rate, the partnership argued (City of Santa Ana 1985b).

In 1989, the estimated $10 million Fiesta Marketplace opened to the acclaim of city officials who said the development was nation's largest master planned, Latin American–style shopping center (Dougherty 1987). There was general consensus among the partners and the broader community about the center's intentions and what it would take for it to be successful. The partners understood that Fiesta Marketplace was a cultural shopping center that would be successful if it appealed to *La Cuatro* consumers. City workers were not projected to be a significant demographic block in sustaining the area. Chase, for example, understood that Fiesta Marketplace was about appealing to and benefitting from *La Cuatro*'s demographic and success. He commented to the *LA Times* in 1985 that he wanted to "create a nice atmosphere for Hispanics, including the undocumented. I hope we never lose that market" (Rose 1985). A few lauded it as a cross-cultural shopping destination. For some in the press, Fiesta Marketplace was intended to be equally attractive to government workers in the nearby offices and Latinas/os living in the surrounding area (Young 1989). Fiesta Marketplace partner Ray Rangel somewhat echoed this perspective with his offering of lower-, middle-, and high-end pricing, the latter allegedly appealing to White middle-class civic employees. Rangel largely generalized this view and his store to the area and told the *Los Angeles Times* that Fiesta Marketplace had two markets: "from orange pickers to office workers who like a good hat and don't mind paying the price for it" (ibid., p. 2).

Fiesta Marketplace's urban design and businesses made it Mexican cultural-themed redevelopment. Architecturally, the center blended colonial and indigenous Mexican elements. Contemporary buildings were adorned with pastel salmon and turquoise tones, colors popular in Mexican colonial architecture. It included a colonial *kiosco* (kiosk), an octagonal structure that is typically used for concerts and cultural celebrations and found in the center of public plazas throughout Mexico and Latin America generally. The area also included a *mercado*, a traditional pre-Columbian public market concept which typically specializes in a mix of affordable goods and options, including pottery, toys, fresh produce, and meats. The promenade also contained a carousel, a ride more common in the United States than Mexico. Fiesta Marketplace's cultural-themed design emphasis not only helped address community concerns over *La Cuatro's* fate, but visually represented a deepened commitment to the area. We can more clearly understand this urban design commitment when we consider that *La Cuatro* was more representative of art deco architecture and neutral colors.

The shopping center was also responsive to sustaining businesses that appealed to the Latina/o community. At the soft grand opening, eighty percent of

Fiesta Marketplace's forty commercial spaces were open to the public, offering affordable cultural products and other goods that were generally unique to the zone, but also in line with *La Cuatro*'s demographic market (Young 1989). For example, sold and unique to Fiesta Marketplace were soccer uniforms and equipment; *ropa, botas, y sombreros de vaquero* (western clothing, boots, and hats); baked goods; and ice cream. The mercado offered traditional Mexican merchandise such as piñatas, huaraches, marionettes, and *ropa de manta*, traditional coarse cotton clothing associated with indigenous and Mexican rural communities (Young 1989). Fiesta Marketplace and *La Cuatro* shared the same target consumers, the Mexican and working-class immigrant, and maintained affordability in the area.

Fiesta Marketplace and the Politics of Economic and Community Development

The politics of redevelopment raise questions about Fiesta Marketplace's commitment to integrating economic self-interests with community equity ideals. These politics inherently involve the city. For example, Londoño and González (2015) argue that the city's discourse and practice of exclusion of Mexican culture and Latina/o consumers in the downtown is captured, contradictorily, in the history of Fiesta Marketplace, specifically the Yost Theater. As discussed in chapter two, the Yost Theater was a cultural and performing arts center that showcased Mexican musical and cinema legends and made downtown the "happening place" for generations of Mexicans in Santa Ana and Orange County. In the mid-1980s, when plans for Fiesta Marketplace were under way, Louie Olivos Sr., the owner of Yost Theater, was one prominent Mexican-American merchant on Fourth Street who was unable to participate in the city's new venture. Though the Yost had done much to consolidate the Mexican consumer on Fourth Street that Fiesta Marketplace was heavily counting on, the city bought out the Yost Theater and sold it to one Fiesta Marketplace partner who had no experience in cultural entertainment, Irving Chase.

Some in the community believed that this transaction between the city and Chase was an ethically dubious transaction and undermined the very same community concerns and aspirations that Fiesta Marketplace and the city's altered discourse allegedly sought to represent (Londoño and González, 2015). For example, during the city's application for the HUD grant, the city told Yost's owner Louie Olivos Sr. that he would have to bring the theater up to seismic code in order to remain at the site. Olivos took out a high-interest loan to comply, but he was unable to make his final payments and the bank threatened to foreclose. The city then assumed the loan for $600,000, less than the value of Olivos's loan, and sold it in 1985 to Irving Chase, for $50,000, even though the Olivos family offered to buy back the theater from the city for the selling price of $600,000 (McLellan 1999, Community Redevelopment Agency of the City of Santa Ana 1985). Supporters of the Olivoses argue that some city officials denied Olivos this opportunity because they were politically connected with some Fiesta Marketplace partners. Louie Olivos, Jr. does not typically implicate Fiesta Marketplace partners in his personal loss, but certainly holds the city accountable. Olivos Jr. told the *Los Angeles Times,* "as long as I live, I'll say that the city stole the theater from us," expressing his devastation at the loss of his family's business and legacy (Schwartz 1989).

Almost immediately, this transaction made obsolete the cultural music entertainment and programming that was significant to the area and the Mexican community. The paradox is apparent when we consider that within a year of Fiesta Marketplace's opening, Chase cut off large amounts of immigrant families and general patrons that had been routinely cherishing the *Cine Yost*, the same demographic block that general Fiesta Marketplace partners were actively trying to attract. The Yost Theater would be closed to the general public for about 25 years, only to be leased occasionally to church groups and intentionally allowed to deteriorate, admitted Chase (Bose 2011). Chase contends that he closed the theater because he was unable to profit on the investment and secure entertainment on a regular basis. Chase added, "When we opened Fiesta, we renovated the Yost. We spent $750,000 renovating it so we could use it for entertainment purposes. But then we could not generate enough activity to keep it open and use it on a regular basis, so it operated as a church for many years, and it was allowed to decay" (ibid., p. 1).

Fiesta Marketplace also represents classic economic development principles that were detached from formal community participation processes and the ongoing community-wide struggles to preserve *La Cuatro*'s identity. For instance, Fiesta Marketplace represented for Ceballos an opportunity to one day send "my children to a university when they grow up." For these partners, in general, Fiesta Marketplace promised to benefit them directly. It remains unclear, however, whether the partners had any role advocating for community-wide participatory processes for Fiesta's development. In my review of Fiesta Marketplace documents, there was no mention of opportunities for community participation involvement for Fiesta Marketplace. The visioning and financial arrangements were largely, if not entirely, based on restricted networked politics among the partners and the city.

Significant about this arrangement too is that community equity ideals stopped with Fiesta Marketplace, on the eastern edge of *La Cuatro*. The Partners stopped short of advocating for similar redevelopment benefits for other parts of *La Cuatro*. This combination of the Fiesta Partners and city planning, though "participatory" with the Fiesta Partners and infused with community equity principles, is significant given the contentious relationship between Mexican residents and merchants over revitalization in the broader La Cuatro area and generous city subsidies that were used for Fiesta Marketplace.

Note

1 The city replied to my Public Records Act Request for this grant application and stated that it does not have a copy. The city did, however, provide a HUD draft grant

application to the Downtown Property Owners and Merchants Working Group in 2007 and a member of the Group provided me a copy of the draft, which I discuss in this chapter.

References

Bose, Lilledeshan. 2011. "The Yost Theater is ready for its closeup." *Orange County Weekly,* July 28, 2011. www.ocweekly.com/2011-07-28/music/the-yost-theater-santa-ana-olivos-chase-family-dennis-lluy/2/.

City of Santa Ana. 1985a. Resolution No. 85-85. Santa Ana.

City of Santa Ana. 1985b. Urban Development Action Grant: Site A-8 Project.

City of Santa Ana. 1985c. Urban Development Action Grant: Site A-8 Project. Draft.

Community Redevelopment Agency of the City of Santa Ana. 1985. Disposition and Redevelopment Agreement by and between Community Redevelopment Agency of the City of Santa Ana and Fiesta Marketplace Partners a Limited Partnership, edited by Community Redevelopment Agency of the City of Santa Ana.

Dodson, Marcida. 1984. "Santa Ana boasts ambitious redevelopment plans." *Los Angeles Times*, September 24, 1984, 3–5.

Dougherty, Kathleen. 1987. "It's time for the Fiesta to begin–construction starts soon on four-block mercado downtown." *The Orange County Register*.

El Quetzal. 1977. "Orange County barrios unite." *El Quetzal*, February, 1977.

Lejano, Raul, and Anne Wessells. 2006. "Community and economic development: Seeking common ground in discourse and in practice." *Urban Studies (Routledge)*. 43 (9):1469. doi: 10.1080/00420980600831684.

Londoño, Johana, and Erualdo R. González. 2015. "The changing politics of Latino consumption: debates related to downtown Santa Ana's new urbanist and creative city redevelopment." In *Race and retail: consumption across the color line*, edited by Mia Bay and Ann Fabian. New Brunswick, NJ: Rutgers University Press.

McLellan, Dennis. 1999. "The show must go on." *Los Angeles Times*, January 3, 1999. http://articles.latimes.com/print/1999/jan/03/news/cl-59846. Accessed June 20, 2015.

Rose, Andy. 1985. "Santa Ana: federal grant completes financing for center." *Los Angeles Times*, October 2. http://articles.latimes.com/1985-10-02/local/me-16095_1_shopping-center. Accessed June 4, 2015.

Schwartz, Bob. 1989. "For some, fiesta is nothing to celebrate." *Los Angeles Times*, January 8, 1989. http://articles.latimes.com/1989-01-08/local/me-496_1_fiesta-marketplace.

Semanario Azteca. 1981a. "Alcalde: vamos a mejorar Sta. Ana: Gorden Bricken conversa con las Organizaciones Comunitarias." *Semanario Azteca*, June 10.

Semanario Azteca. 1981b. "Mas construcciones en Santa Ana: nuevas viviendas y oficinas." *Semenario Azteca*, December 2, 1.

Semanario Azteca. 1982a. "Habla A.J Wilson City Manager de Santa Ana." *Semanario Azteca*, January 13.

Semanario Azteca. 1982b. "No al plan de desarollo manifiesta la comunidad." *Semanario Azteca*, June 9, pp. 1, 7.

Semanario Azteca. 1982c. "No al plan de desarrollo dicen sectores vecinales de Santa Ana." *Semanario Azteca*, May 19, pp. 1,10,11.

Semanario Azteca. 1984a. "Amargura en Santa Ana." *Semanario Azteca*, June 27, 1,4.

Semanario Azteca. 1984b. "Molestia en comunidad Latina ante accion discriminatoria." *Semanario Azteca*, April 25, 1,9.

Semanario Azteca. 1984c. "Santa Ana decide otorgar ayuda." *Semanario Azteca*, May 23, p. 1.

Semanario Azteca. 1985. "Controversia en Caso Pulido." *Semanario Azteca*, April 10, pp. 1, 11.

Snow, Anita. 1987a. "Santa Ana: a city in transition to become county hub once more." *Orange County Register*, June 7, 1987.

Snow, Anita. 1987b. "Winners and losers: change brings tears, cheers." *Orange County Register*, June 7, 1987.

Young, Karen Newell. 1989. "*Fiesta Marketplace in Santa Ana offers cross-cultural shopping*." *Los Angeles Times*, April 21, 1989. http://articles.latimes.com/1989-04-21/news/li-2434_1_fiesta-marketplace-ice-cream-shop-custom-auto-service. Accessed February 20, 2014.

3 *La Cuatro* Under Threat, 1990–2010s

A critical issue confronting local urban redevelopment policy-makers in the 1990s and early 2000s was determining whether planning for and marketing to Mexican working-class immigrants should influence the future of the downtown area, or rather if the interests of creative-class and middle-class consumers and residents take precedence. The city's latest attempt to implement new urbanism (NU), creative cities (CC), and transit-oriented development (TOD) strategies of urban redevelopment exacerbated much of the controversies from past decades, but also created new tensions. While proponents of the NU, CC, and TOD approaches to planning praise diversity and inclusiveness, the plans themselves are consistently linked to gentrification and the displacement of marginalized people from cities (Diaz 2012, Peck 2005, Londoño and González 2015). Indeed, planners often chart growth and change away from the existing and majority community and toward an imagined community. As Holston argued, modern planning seeks "to transform an unwanted present by means of an imagined future" (Holston 1989).

In this chapter, I ask: how do the discourse and practice of downtown redevelopment work to construct the imagined community? How does this erase the existing majority community from consideration? What are some practices to achieve these aims? Last, what does the downtown district look like during the transition from serving Mexican, working-class, immigrant consumers to serving CC and middle-class consumers? Figure 3.1 is a map of the downtown area and two major redevelopment projects that are the focus of this chapter—Artists Village and East End and the area's largest and busiest zone, *La Cuatro*.

The Changing Urban Demographic Context

In 2010, Santa Ana was California's eleventh-largest city, a figure not too surprising given that the city has historically been Orange County's most populated city, only recently losing that title in this same year (U.S. Census Bureau 2011). We can appreciate the magnitude of this city population size by considering the following. In 2010 Santa Ana had twice as many people than it did in 1970, growing from 156,983 people to 324,528 people. The racial and ethnic demographics had also shifted substantially in 2010. The percentage of people identifying themselves

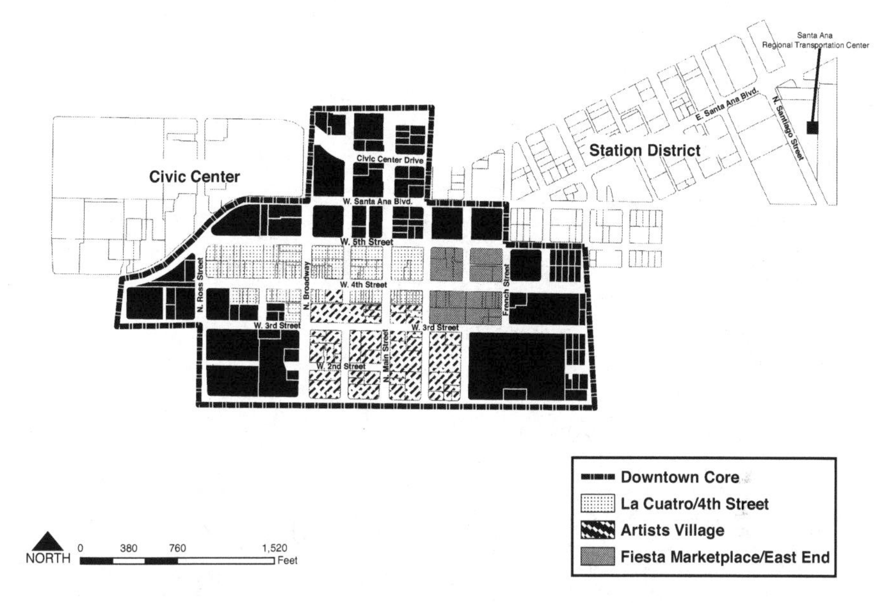

Figure 3.1 Map of downtown redevelopment projects.

Source: U.S. Census Bureau.

as White withered to 9 percent, a significant reduction in this population given that the mark was 71 percent in 1970. However, the Latina/o population grew from 22 percent in 1970 to 78 percent in 2010, distinguishing the city with the seventh highest percentage of Latinas/os in the United States among places of 100,000 people or more (U.S. Census Bureau 2010). Residents of Mexican origin made up over 70 percent of Latinas/os in the city. Latinas/os of Central American descent were the second most visible group. Asians were 11 percent, Whites 9 percent, and African-Americans 2 percent of the population (ibid.).

By 2010, Santa Ana's immigrant population had skyrocketed. The immigrant population at this time was roughly 11 times larger than it was in 1970, accelerating from 13,918 people to 160,779. This means that about half (49 percent) of the city's more than 300,000 population was immigrant. This shift toward an immigrant city was so significant that nationally, Santa Ana in 2010 had the nation's second-highest proportion of immigrant residents for large cities (U.S. Census Bureau 2010, Figure 3.2).

With population growth and change also came poverty and the many problems associated with it. In 2004, the Nelson A. Rockefeller Institute of Government's national Urban Hardship Index ranked Santa Ana number one in the U.S. for unemployment, the number of dependents, education, income, crowded housing, and poverty (Montiel, Nathan, and Wright 2004). In 2009, Santa Ana reported that families of four were mostly working class and poor—60 percent were in the

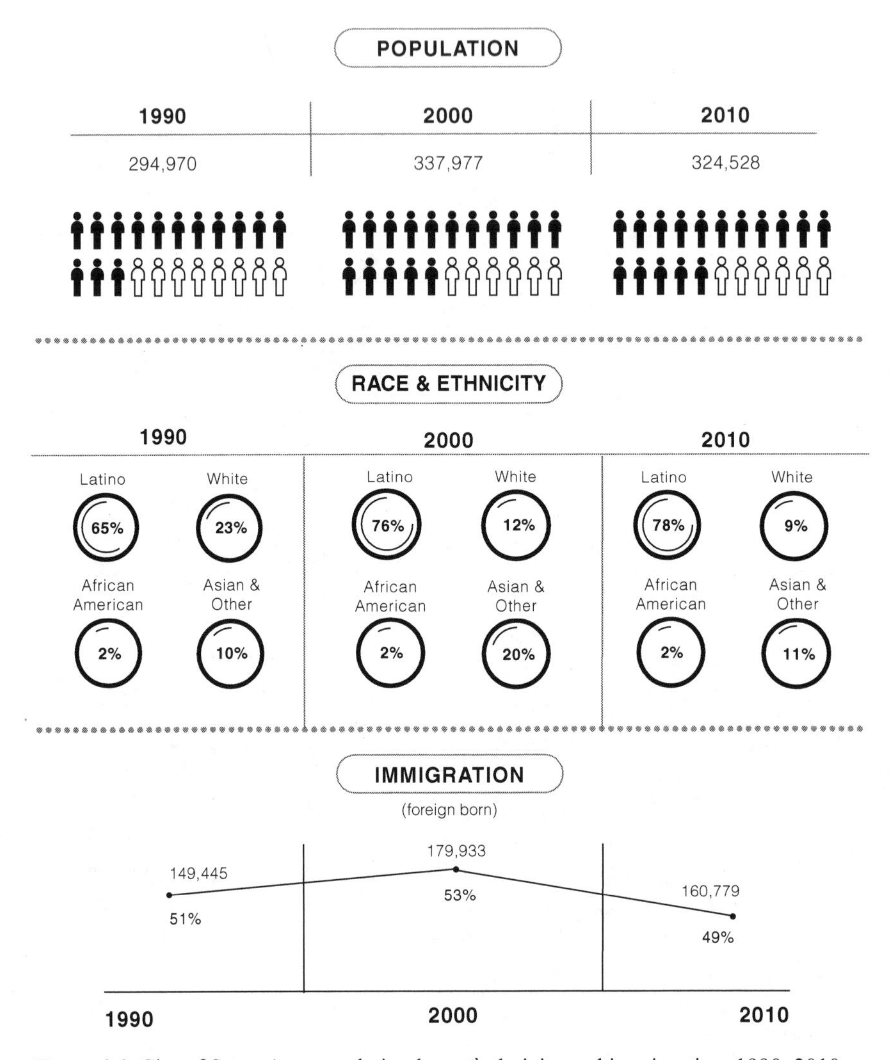

Figure 3.2 City of Santa Ana population by rac`ethnicity and immigration, 1990–2010.

Source: Data from U.S. Census Bureau, 2011. American Fact Finder, Selected Social Social Characteristics in the United States, 2011 American Community Survey 1-Year Estimates, Santa Ana, California; U.S. Census, 2007. SOCDS Census Data Output: Santa Ana City, CA. Figure created by Cara Gafner.

"low" and "extremely low" range of income, earning less than $74,400 (City of Santa Ana 2009). The 2009–2013 American Community Survey estimated that 21.5 percent of persons in Santa Ana lived below the poverty level and the per capita income was $16,374, less than half of that of Orange County as a whole. These income levels certainly make it difficult for most residents to own homes or have supplemental income beyond that for bare essentials. For instance, approximately 50 percent of residents are renters and about half have housing costs that exceed 30 percent of their income (City of Santa Ana 2009, The Planning Center 2008). With this level of poverty the city can be expected to accommodate significant areas suffering from problems of urban deprivation.

New Urbanist, Creative City, and Transit-Oriented Redevelopment

The ten-block Artists Village, Orange County's first publicly subsidized artist colony, opened on the southwest portion of the downtown and within strolling distance of *La Cuatro* in 1994. It consists of historic buildings, anchored by the 1929 Spanish Colonial revival Santora Building (Figure 3.3), which features an elegant gated lobby entrance, and the brick Italian Renaissance Revival 1924 Grand Central Building. The Santora Building houses a few restaurants and most of the galleries in the Artists Village. The Grand Central Building is a half-city block deep and full city-block long, two-level, mixed residential, commercial, and educational complex. The Grand Central Building includes 28 studio apartments strictly for students of the university for which I work. The Center services graduates and in-residence artists, and has galleries, classrooms, and a theatre and café. The Second Street Promenade is the Village's central courtyard; it connects both buildings. Equal in length to the Central Building, the courtyard has lights in the trees for visitors to enjoy while among the now 40-plus galleries, art studios, retailers, cafés, and late-closing restaurants, along with bars in the area that primarily service the nighttime consumer, and a monthly Art Walk. The Orange County Center for Contemporary Art, a nonprofit housed in a former body shop across the street from the Grand Central Building, "is dedicated to the pursuit of professional excellence and freedom of expression in the arts" (Orange County Center for Contemporary Art (OCCCA) 2016). The Village also consists of housing typical of new urbanist downtown redevelopment in the city. This includes about 100 work/live loft units and apartment homes, which contain the Santa Ana Lofts, Main Street Lofts, and East Village Lofts and The Court at Artists Village.

This Artists Village idea is said to have begun with Don Cribb, a Santa Ana native who lobbied the city for art redevelopment. City officials and artists that Cribb mobilized typically coined him "the prime mover" and "brainchild" behind the Artists Village's arts movement in Santa Ana. He began "fervently" fighting behind the scenes during the late 1980s to gain community allies and city collaboration for his vision (Roberts 1998, p. 60). He reached out to City Council members, community development staff, residents, and community groups. He went on to found and become president of the Santa Ana Council of Arts and Culture (SACAC) in 1990 and carried out small-scale arts projects with people who supported his cause and were willing to rally around the arts as a redevelopment strategy, such as the Santa Ana Historical Preservation Society. SACAC formed with 14 members, "and throughout that lobbied for community development block

Figure 3.3 The Santora Building at the Artists Village in 2008.
Source: Author's personal collection.

grant funds," Cribb stated (ibid., 58). As Cribb points out, he leveraged SACAC to successfully steer the city to commit subsidies and "to assist with the donation of land and exclusive right-to-negotiate guarantees" (Marsh 1994, p. 163).

The city invested substantially in the Village. The City Council approved $5 million in federal community development block grants and $1 million in general redevelopment dollars. Most of the funds were used by the city to purchase and renovate buildings and to provide forgivable loans for rehabilitation. For example, the City renovated the former 6,300 square-foot body shop for OCCCA and granted title to the nonprofit, an organization of which Cribb was a board member (Mena 2002, Gurza 1995). The city leased the Grand Central Building for 10 years at $1 per year to CSUF's Art Department. Both centers control, manage, and sublease to self-sustain. The city also provided a forgivable loan of $500,000 to then Santora owner/area investor Mike Harrah to remodel.

The Artists Village was not very successful in attracting foot traffic at first, but within ten years of its opening, persistent programming and more support from the city for some businesses changed that. From the grand opening in 1994 to the summer of 1995, the first dozen artists opened galleries in the Santora Building, a number much lower than anticipated. Six years after the first artist moved in to the Artists Village, the *Orange County Register* reported that the zone had disappointed both local artists and politicians (Garlington 2000). Studio gallery owners noted that there were not enough customers, rents were too high, repeat business was low, and Monday through Friday foot traffic was subpar. One of the Artists Village's strongest supporters, Councilmember Lutz, acknowledged that the area lacked a "critical mass" (ibid., p. 3). Coordinated attempts by Artists Village businesses and leaders had not attracted the foot traffic and consumers they hoped for.

While the Artists Village continued to generally struggle into the mid-2000s, some local publications were reporting signs of progress. "I am happy to report that the First Saturday Open House is still a lively affair," said Richard Chang of the *Orange County Register* (Chang 2006). Computer- and technology-based art, interactive video, and more traditional art galleries helped draw larger crowds. Bars, nightclubs, and restaurants began to move in and evening foot traffic grew on Saturdays.

The city helped with this trend. The city became more flexible with liquor licensing and allowing drinking establishments. For example, the city had long limited alcohol licenses downtown, and had razed several *cantinas* (bars) on *La Cuatro*. Proof Bar, however, obtained a Type 48 liquor license in 2006, gaining the right to sell hard liquor until 2 am. Bar owner Joey Mendes, speaking to *O.C. Weekly*, acknowledged the city's longstanding downtown alcohol license suspension and his privilege, calling his license a "rarity and has not been granted in the city since 1986" (Segal 2007). Other drinking establishments followed.

By the summer of 2011, *The Voice of O.C.* reported that the Village was doing quite well as a whole. Restaurants in the Village were reporting double-digit sales increases from previous years, though it is unclear how low or high previous years were, the local parking lot operator was hiring more staff for the once a month Art Walk (formerly called Open House) than in previous months, and bar bouncers were generally seeing identification cards (IDs) from all over Orange County (Elmahrek 2011a).

While the Artists Village had a slow start and some signs of general success, the city nevertheless had moved quickly in preparing a similar and more ambitious downtown redevelopment initiative. The Renaissance Specific Plan (RP), which "envisions instant gentrification," and the Station District (SD) within it, would be the city's largest and most ambitious initiative, combining creative city, new urbanist, and transit-oriented principles (Delson 2008b).

The Santa Ana Renaissance Plan and Station District

The city released the RP in late 2007. The most comprehensive rezoning plan in the city's history, the plan aimed to provide the city and developers a framework to revitalize 421 acres over 124 blocks, stretching one mile from the county's civic center to the city's train station. The plan was crafted to redevelop the downtown district, low-density pre-World-War-II housing; two barrios (labeled in the plan as "diverse neighborhoods"); a train station center; a regional employment center containing local, state, and federal buildings; and a major street. Embedded among these plan areas were industrial facilities, city-owned boarded-up homes, aging apartments, poverty, and other urban issues (Delson 2008b). The plan's transit-oriented emphasis sought to strengthen the city's role as a regionally important transit-oriented district and government center (Moule and Polyzoides Architects and Urbanists 2007).

The RP conforms to the NU design style and rhetoric. The focus is mixed-use, mixed-income, and pedestrian-friendly spaces. The city's objective is to address historic preservation, mobility and public transportation, social and cultural considerations, open and green spaces, housing, and employment. It also seeks to imitate "traditional" American neighborhoods:

> The plan works in every way to recognize and enable traditional neighborhood development of varying intensities, including transient-oriented and commercial districts, through a tailored vision, policies, and regulations. The plan is based on a set of integrated principles that have produced the best places and cities throughout the world. (Moule and Polyzoides 2007, 1:7)

The plan promoted a number of health and social benefits. Four thousand housing units, including lofts and live-work units, would encourage living above stores, thus reducing use of automobiles, increasing access to local amenities, and providing "diverse opportunities for a variety of incomes" (Moule and Polyzoides 2007, 1:8). A mix of architectural buildings would enhance the character of the area, the plan added. Retail would encourage pedestrian storefront shopping and accommodate cars. Street design would diffuse traffic, encourage people to walk and bicycle, and improve pedestrian safety (ibid., 1:8). According to the RP, Santa Ana was ready for a new redevelopment cycle and could build on a long and distinguished history with existing architecture and land (ibid.).

To prepare the RP, the city hired consultants Moule and Polyzoides Architects and Urbanists, a firm whose part owner helped found the Congress of New Urbanism (CNU). Per the RP, the plan was partly prepared with feedback from a cross section of community participants who engaged in visioning exercises. The plan describes the participants as downtown merchant groups, historic-preservation groups, neighborhood representatives, industrial business owners, and residents generally. These participants engaged in community interview workshops, study sessions, a six-day charrette process, a series of focus sessions, and other large visible processes (Moule and Polyzoides 2007, 1:6).

Perhaps in part due to the vocal opposition to the RP (discussed in the next chapter), the city decided to repackage a portion of the RP into a proposed 94-acre SD, focused around a new Transit Zoning Code (González and Lejano 2009). Creating a new transit code involved rewriting the Transit Zoning Code to allow for the SD project. The city touted the SD as the key to creating a new transit-oriented city and county, connecting Santa Ana via transit to the region.

The Go Local streetcar, a 4.2-mile system, is the SD's focal point. In addition to the RP's five main areas for redevelopment and its connection to major freeways, the SD includes the French Park neighborhood, a middle-class neighborhood known for its Victorian homes, the Artists Village, county property, and approximately 50 parcels the Agency and City own along a major corridor.

Some local activists pointed to the SD as essentially the same as the original RP, but with some zoning modifications. The form-based code designations, such as "Urban Neighborhood" and "Urban Center," are the same in the RP and SD, with some small differences. One minor difference is that the SD refers to the Rail Station as a Transit Village.

In reality, the SD represented the city's longstanding vision to become a transit-oriented downtown, a vision preceding the RP. The city pursued the SD having already made a commitment to a failed regional light rail project and related redevelopment projects about three years prior to the RP. As an example, the Redevelopment Agency had begun utilizing State of California 20 Percent Set Aside Funds for years before the RP and by 2009 had purchased fifty-five residential properties across 11.76 acres in the RP area (City of Santa Ana 2007, The Kennedy Commission 2008). According to the city, property owners sold voluntarily and this allowed it to absorb the majority of properties from the failed regional light rail project. The Santiago Street Lofts project that opened in 2006 was another sign that the city had committed to TOD prior to the RP. The lofts are situated across the street from the Santa Ana Regional Transportation Center offering commuter trains to San Diego and Los Angeles. The city touted the Santiago Art District and Transportation Center as the gateway to downtown.

Making way for redevelopment required buying out residents and creating blight, most of this activity in the Lacy barrio. As an example, most of the residential properties that the Redevelopment Agency purchased and boarded or leveled were in Lacy (City of Santa Ana 2007, The Kennedy Commission 2008). The barrio consists of about 35 blocks and 7,000 residents, of which over 90 percent are Latina/o, the majority of whom are Mexican immigrants and Mexican-American (City of Santa Ana 2008, 2007, U.S. Census Bureau 2000). Households in the area include, on average, about 5 persons; 86 percent of these households are renters and 90 percent speak Spanish at home (U.S. Census Bureau 2000).

Redevelopment Prompts Protest

Outbursts and battles stretched and intensified across time and space—from the 1990s Artists Village to the Station District, and these intensified in the early 2000s. Initially, these protests pointed to racial and class issues associated with interpretations of merits and alienating consequences. Direct grassroots activist challenge to this redevelopment, however, was virtually non-existent. In this respect, it appeared within the broader community as though the City Council had found redevelopment models that spoke for the city majority and spent public money in line with the community's priorities.

Councilmembers were overall supportive of the Artists Village. While the city's generous financial investment in the project is perhaps the clearest sign of support, public endorsements are other indicators of approval. For example, in the mid-1990s, councilmember Lutz commented that the historic architecture in the area made a natural case for the Artists Village. For him, the downtown's art deco urban design was compatible with creative-city redevelopment ideas that glorify historic architecture and an obvious choice for the city. Councilmember Lisa Mills showed her approval by offering an optimistic city revival prediction 16 years into the future, giving the Artists Village full credit for making Santa Ana's central core the cultural destination for all Orange County by what would be 2010. "Santa Ana is the cultural hub of Orange County ... The City's Artist[s] Village is the number one night spot in Orange County and the demand for retail space is so high that the City will expand the village boundaries again ... and there is a waiting list ... to buy a house in many of the City's unique and charming neighborhoods," she exclaimed (Marsh 1994, p. 161). Lutz and Mills, as well as some other councilmembers, had become the latest city councilmembers manifesting institutional memory, representing a long line of city officials giving legitimacy to redevelopment heavily, if not exclusively, focused on appealing to and serving those outside the city.

There was some small but vocal opposition to the project a few years from the inauguration, however, though most was from a few city officials. As *Orange Coast* magazine put it, "debates raged, and passions were intense" over the Artists Village (Gurza 2009, p. 84). For example, three years after the Village opened, councilmember Ted Moreno accused fellow councilmembers of supporting the development to satisfy their personal interests and that of city middle-class professionals, calling the Village a "hangout" for council members' "friends" (Dubin 1997). Speaking to *Orange Coast* magazine, the councilmember voiced his opposition to ongoing city funding for the Village, saying funds were not well spent on city priorities:

> The taxpayers' money should be spent to improve infrastructure and on the parks system to improve things for our children. We've seen what happens in this city when children aren't looked out for—they get into gangs. (Roberts, 1998, p. 63)

Subsequent to the councilmember's remarks, Santa Ana Planning Commissioner Sean H. Mills took to the newspaper media and accused city actors and allied activists in the wider policy and political processes driving elite redevelopment and gentrification, the latter a term that was likely used for the first time among local community activists, let alone city representatives. In his opinion piece to the *Los Angeles Times*, Mills stated "And now that this tax-payer subsidized pipe-dream of center founder Don Cribb and his cronies on the City Council is coming to fruition, they want to discriminate against businesses they deem not upscale enough." Like the councilmember, he cited broader public works problems in the city that he felt go virtually neglected, adding "I don't understand how we can afford to pay for the 'artists village.' I say let the art lovers pay for it." (Mills 1999). He took to the media once again five years later. In his opinion piece titled "Ethnic Cleansing in Santa Ana?" for the *Orange County Register*, Mills expressed disbelief and his tone was angry, writing that narrow political interests backed a project that he believed would gentrify the majority Mexican and working-class families in and around the downtown. Mills wrote:

> Suggestions that projects such as the 37 story office tower, pricey artists lofts or the "Artist[s] Village" in general signal rebirth of downtown Santa Ana are absurd and insulting. The community consists mostly of immigrants and working-class families of modest economic means who fail to benefit from the developments. What we see is not revitalization, it is gentrification. (Mills 2004)

Mills also accused city officials of socially constructing downtown economic problems and favoring a different and bourgeois community, noting, "rather than build on the success of the bustling business climate of the downtown area, where small family-owned ventures cater to an ethnically diverse consumer base, they have chosen to use the politics of greed, elitism and division and work against it" (ibid.).

A *Los Angeles Times* article by Gustavo Arellano (2007a) made similar claims and mocked general perceptions of the success of the Artists Village Art Walk sometimes reported in mainstream media by comparing it to a highly successful non-Artists Village Mexican cultural celebration downtown. He wrote:

> Instead of investing money in more police officers, Santa Ana spent a bit more than $1 million to refurbish and repaint a water tower so it could read "Downtown Orange County." Rather than improving infrastructure, it installed a downtown water fountain better suited for Tivoli. Instead of supporting local businesses, Santa Ana lures such out-of-town corporations as Starbucks and American Apparel with massive subsidies. And the city's primary project over the last decade has been an Artists Village that gets its largest crowds not during its "first Saturday" gallery open houses but during Día de los Muertos. (Arellano 2007a)

With this last point, Arellano was mocking the Village's self-perception that it hosts successful Art Walks by comparing it to the *Día de Los Muertos Noche de Altares* also held downtown. The largest all-day November cultural celebration regularly hosts about 100 altars honoring the deceased and critically depicting societal issues. It is held along *La Cuatro* and outside of the Artists Village boundaries and organized by the nonprofit El Centro Cultural de México and Calacas Café. The celebration is also one of Southern California's largest yearly Mexican-indigenous cultural celebrations. In 2014, for example, the cultural celebration attracted 30,000, say committee organizers and the Santa Ana Police Department.

The RP only amplified racial and class controversy over redevelopment. Some touted the plan a good response to perceived downtown deterioration and a path to increasing the downtown's tax base. "Readers' comments" sections of area newspapers and Internet blogs from Orange County recorded routine discontent over downtown Santa Ana's high concentration of consumer options marketing to Mexican immigrants (Arellano 2007b, 2008b). To some property owners, politicians, and business leaders, the downtown area could improve its image and tax revenue by attracting a wider market beyond Latinas/os and working-class immigrants (Delson 2008b, Taxin 2006).

La Cuatro merchants had specific problems with the RP. Some *La Cuatro* merchants questioned the need for the RP, arguing that it would squeeze out the current businesses and customer base, instead of enhancing an already thriving district, a gentrification trend also happening in other cities in Southern California (Becerra 2008, Community Based Coalition Addressing the Renaissance Plan 2007, Taxin 2006). In an independent survey conducted by a local university master's student, another group of downtown business owners ranked fear of change and concerns about the loss of the area's Latina/o character second among their top five planning concerns, next to parking and traffic (Thakur 2008).

Some business owners expressed frank concern that the city wanted a more upscale and white population (Community Based Coalition Addressing the Renaissance Plan 2007). In a two-page letter to the City Council, a 14-member ad-hoc group of property owners and merchants along *La Cuatro* requested to collaborate in the drafting of the RP before it went to City Council to vote because they were "seriously concerned that almost no input was received from the property owners and merchants of the very dynamic and economically productive downtown commercial district" (Downtown Property Owners and Merchants Working Group 2007, p. 1). The group pointed out that their overarching goal was "for redevelopment that could serve local needs (in Santa Ana this is a Latino market) while at the same time welcoming and serving neighboring communities, downtown's daytime population and the increasing purchasing power and discerning spending of the Latino consumer" (ibid., p. 2).

City supporters of NU, CC, and TOD tended to deny any complicity in the population changes that transpired during these types of redevelopment, instead suggesting that city demographic changes compelled the city for such redevelopment. For example, the U.S. Census shows that the downtown area's two census tracts lost an average of 13.65 percent of their residents from 2000 to 2010. While the downtown area became less dense, the racial composition also changed; whites increased by 17.4 percent and Latinas/os decreased by 16.33 percent. We can understand that redevelopment in the downtown helped with these trends when we consider that during the same period citywide, the same census reveals about a 4 percent population decline and a slight increase (0.04 percent) in the Latina/o population. The most drastic demographic changes in the city were in the downtown area. Instead, city supporters suggest that these demographic shifts point to changes in consumer desires and residential patterns. National and local political and economic conditions, such as high rates of foreclosure and the Obama administration's draconian immigration laws, have clearly contributed to some population decline in Santa Ana.

But, local anti-gentrification advocates and many Latina/o longtime business owners conclude that downtown redevelopment is overwhelmingly responsible for these downtown demographic changes. At the same time, these stakeholders insist that a minor drop in the Latina/o downtown population and the national recession did not justify redevelopment projects that dismiss the still majority and large number of Latina/o downtown consumers and the city's majority Latina/o community in favor of a reimagined consumer base.

Spanish- and English-language television, print, and radio news media reported many of these controversies—gentrification, displacement, affordable housing, erosion of Latina/o commercial identity by Latina/o elected leaders, City Council misappropriation of community development funds, inadequate public participation, and urban governance (Greenhut 2007, 2008, Arellano 2008c, Delson 2008a, b, Parsons 2008, Gurza 2009, Irving 2010, 2011, Mantle and Turac 2011). *Telemundo*, an American Spanish-language broadcast television network, featured the Renaissance Plan and titled their story, "¿Desalojo Etnico?" ("Ethnic Displacement?"). A Southern California Public Radio story implicated the all-Latina/o city council in these downtown redevelopment debates with the headline, "Will Latino-led redevelopment in Santa Ana ultimately rob it of its Latino-ness?" (Rojas 2011). *The New York Times* story was headlined "In Santa Ana, new faces and a contentious revival" (Medina 2011). Residents' criticism of the plan's claim to "rebirth" the neighborhoods and the downtown occurred in this context.

Racial and Ethnic Representation in the Historic All-Latina/o City Council

In 2006, three Latina/o candidates won seats on the city's seven-member council, joining four Latina/o incumbents, and Santa Ana had an all-Latina/o City Council for the first time—the first U.S. city with a population over 300,000 to have an all-Latina/o City Council (Delson 2006). It was a turning point in racial and ethnic redevelopment representation politics. Nonetheless, in 2006, redevelopment controversies persisted, testing the assumption of the racial and ethnic representation model. The City Council's racial and ethnic composition was indeed historic, both locally and nationally. The assumption locally followed the racial and ethnic representation model: that the all-Latina/o city council would be beneficial to the Latina/o group and city because the co-ethnic official would likely have the disposition to better understand and attend to their interests than members of other groups (Thomas 2008).

Questions in the city surfaced about the extent that the racial and ethnic identity of the all-Latina/o City Council could translate to socioeconomic interests of the large poor and working-class Latina/o population in the downtown area. As we learned in previous chapters, activist support for racial and ethnic representation in the City Council and city generally tended to coincide with advocacy in defense of the city's predominant and growing working-class and Chicano and immigrant Mexican populations. This council was a noteworthy change from the City Councils of previous generations when a majority of white men persistently occupied those seats and Latinas/os were heavily underrepresented. "A generation ago, Latinos complained of an all-white City Council. After becoming Santa Ana's dominant ethnic group in the 1980s, they agitated for better representation—a push that eventually helped elect Pulido, among others,",pointed out a news story in the *Los Angeles Times* (Carcamo 2015).

Some City Council members' perspectives highlight the fallibility of the racial and ethnic model of representation that translates to socioeconomic interests of the large poor and working class (Londoño and González 2015). Councilmember Bustamante was born and raised in Santa Ana and is the son of Mexican immigrants. In spring 2011, Southern California public radio station KPCC's daily talk show *AirTalk* set up a panel in Santa Ana to debate the pros and cons of the city's redevelopment. The panel consisted mainly of members from the Santa Ana Collaborative for Responsible Development (SACReD), a grassroots group who opposed the RP and the SD, and Councilmembers Bustamante and Martinez. Bustamante spoke indirectly to activists and general supporters for sustaining and enhancing *La Cuatro*'s identity. He commented that the downtown is not surviving with the existing business base and redevelopment is helping turn this around. He extolled, "I really think it's [Downtown Santa Ana] experiencing a rebirth, a renaissance" and said if you do not redevelop, "you either move with it or die in line" (ibid.).

Bustamante also labeled *La Cuatro* exclusionary of higher economic strata within the city, a strata he openly identified with and wished to see served. To frame his perspective, Bustamante responded to SACReD panelists and other anti-gentrification advocates in general by asking: "Who are they trying to keep out?" He followed that *La Cuatro* should broaden its market to higher socioeconomic strata, not a different ethnic base: "There should be options for everybody here. The city is not changing ethnically; it's changing socioeconomically." This class appeal is personal. He noted that he does not "want to go someplace else to buy my suits" (ibid.). The class appeal also shows that unlike past generations of city attempts to attract higher class groups from the county, his rhetoric focused within the city. His stance was for redevelopment that could address the city's Latina/o middle class. Santa Ana has a small percentage of middle-class Latinas/os, including the 5.6 percent of city government workers and others who work in related sectors (Londoño and González 2015).

Councilmember Michele Martinez, a fourth-generation Mexican-American who represents part of the downtown area, has emphasized that niche immigrant spaces in the downtown are exclusionary of multiple generations within the city. In remarks to the *Los Angeles Times*, Martinez pointed out that the redevelopment in downtown should not be perceived as against racial or ethnic interests or lack of Mexican pride, but as a sign of city responsiveness to broader consumer interests within the city. She noted: "It is just that not all Latinos are immigrants." (Becerra 2008). She added, "I have nothing against 50 quinceañera shops, but I don't shop there. Many of my friends don't shop there. Parents and grandparents may shop there, but young kids are not going to shop there, unless they're immigrants." (ibid., p. 2). Quinceañera is a Latin American religious tradition of celebrating the fifteenth birthday of teenage girls; Martinez referred to stores that sell bridal wear as well as quinceañera apparel. While she used hyperbole to express her ire toward such businesses (by my own research, there were 15 in 2011), her public comments stopped short of recommending or endorsing actions that would help phase out these or patterns of businesses in the area targeting immigrants. Quinceañera businesses in the area are typically hybrid in that they are also bridal stores (see Figure 3.4).

Bustamante and Martinez do not make the argument that the entire ethnic identity of *La Cuatro* should change or that redevelopment should target an economically better-off regional or creative city demographic. Instead, they primarily craft their objection to the consumer options around economic class and generational difference within the city's Latina/o community. They argue that downtown will not sustain with the existing business base and must attract a market beyond poor and working-class Latina/o immigrants. The city leaders desire consumers with deeper pockets who represent multiple generations outside of the immigrants within the city.

This contrasts the dominant city discourse emphasizing redevelopment targeting the middle class and creative class from the county. While a desire for increased revenue is part of urban policy and Bustamante and Martinez focus on the city and not the county, Bustamante's and Martinez's discourse shares some commonality with the traditional redevelopment discourse that undermines the presence of *La Cuatro*'s consumer demographic and businesses that cater to it.

Many local advocates in Santa Ana were hopeful that the council would end a history of racism and policy discrimination by past City Council administrations and other public officials in downtown commercial development and housing. Yet, it has become obvious that ethnic identity does not evenly translate to socioeconomic

Figure 3.4 Pasarela Bridal.

Source: Author's personal collection.

interests that could support the needs of the large poor and working-class Latina/o population in the downtown area. As member Albert Castillo stated during a SACReD meeting, grassroots campaigns to support a Latina/o candidate are no longer enough; nowadays "it's getting in the right Latina/o" (personal communication). In the city, support for Latina/o City Council members sometimes coincides with support for the predominant working-class and immigrant Mexican population of the city, but at other times the economic interests of City Council members parallel those who are economically better off. This class difference, as we have seen, has led some community activists, residents, and business stakeholders to challenge these Latina/o local power brokers.

Not all Latina/o councilmembers, however, were in favor of gentrification. Claudia Alvarez, a councilmember since 2000 for the ward that includes east Fourth Street, a deputy district attorney, and an immigrant, opposes some of the redevelopment even though most of the council supports it, including the Fiesta Marketplace–East End change. In August 2010, when protests were under way against the newly formed downtown Community Management District, known locally as the Property Based Improvement District (PBID), Alvarez added fuel to the fire by saying at a City Council meeting that Fiesta Marketplace partner Chase was kicking out long-term tenants in favor of hipster people (Londoño and González 2015).

Evoking Mills's earlier letter to the *O.C. Register* and the slogan of some SACReD members, Alvarez pointed out that "Ryan and Irv Chase are very much interested in continuing the ethnic cleansing that is going on downtown" (Galvin 2011). In an effort to reveal the PBID's bias against *La Cuatro*'s main business profile, Alvarez announced at the meeting that when she asked why Downtown Inc. has its offices in a building owned by Chase, she was told the organization had gotten a "great deal" on the rent. She then stated: "So if Hitler rents you a place, he's giving us a great deal, so who cares what he stands for?" (Londoño and González 2015). The Chases, who are Jewish, understood the statement as a personal attack, as did city councilor Bustamante and many other Santa Ana community members who called for Alvarez's resignation. Soon the Anti-Defamation League was pressuring Alvarez to apologize for her anti-Semitic comments, but she skirted apologetic statements, claiming: "I'm not calling him Hitler … I used it as an example. Maybe an extreme example, but if the shoe fits…." (ibid., p. 1).

Other City Council members have largely avoided making public comments about their position on downtown redevelopment, presumably to prevent their personal or their family members' private businesses from being embroiled in the debate (Arellano 2007b). Councilmember Sarmiento has family members who own Festival Hall, a Mexican nightclub and banquet hall that had been removed from the original RP boundaries to escape the RP's regulations and standards. The move, according to the city, was to avoid a conflict of interest, thereby allowing Sarmiento to vote on the RP. Likewise, Ace Muffler Shop in the downtown area (discussed in Chapter 2), owned by Mayor Miguel Pulido's father, was also excluded from the RP boundaries to allow the Mayor to vote on the project without violating conflict of interest laws (ibid.).

As I have shown here and have written elsewhere (Londoño and González 2015), the varied involvement of Santa Ana councilmembers, business owners, and community leaders in redevelopment debates reveals a differentiated Latina/o population in Santa Ana with diverse consumer needs and redevelopment interests that may contrast the interest of poor, working-class, and immigrant Latinas/os. Those who seem to attend to the interests of the latter population invoke race and ethnicity as a salient issue in current redevelopment and believe that the underlying political economic logic of gentrification is a threat to racial and ethnic community sustainability. The urban elites who are most interested in underscoring class and generational differences in consumer preferences see racial and ethnic affinities as exclusionary and little wrong with the current downtown redevelopment. Redevelopment discourse among Latina/o city officials is divided along class and racial and ethnic lines of redevelopment discourse.

Reimagining through Discourse

I now examine how the social construction of the imagined community occurred in Santa Ana, from the Artists Village up to and surrounding the proposed SD. First the planners detailed the ideal types that would make up and benefit from the Artists Village and disparaged the majority consumer and resident constituency in the area. As an example, speaking about the project's formal planning phase, activist Cribb proudly commented that he greatly influenced what he called "imagineering"—imagining and plotting to scheme via redevelopment planning particular beneficiaries—professionals, interior designers, advertisers, architects, art consultants, private dealers, and "creative people" who may want studios (Marsh 1994, p. 164). The city bought into his reimagining. As an example, the city commissioned Robert Charles Lesser and Co., a West Los Angeles real estate consulting firm, in 1993 to study the feasibility for revitalizing the downtown through the arts and attracting the creative class.

Adding value to the creative class, Cribb tags it authentic and describes it this way: "If we could establish an atmosphere for the artists, where the real blood of the creative world flows, we would develop an interactive space with a legitimate heartbeat" (Marsh 1994).

But, justifying reimagination for a particular community could gain some legitimacy by disparaging another who stands in its way. These twin processes, as we learned in previous chapters, were not unique to the Artists Village era. For instance, during the 1970s and 1980s, while Cribb was in his 20s and 30s, he left Santa Ana and traveled to New York, San Francisco, London, and parts of Latin America and Europe, among other places, and became increasingly fixated on replicating artist districts, such as SoHo, in Santa Ana. He moved back to Santa Ana in 1986 convinced that arts redevelopment in the downtown area could remedy his reviled years in the city during young adulthood. "I hated growing up here," said Cribb (Gurza 1995). He had disdain for what he felt had taken over the downtown and city he grew up in, "overcrowding, graffiti, crime, and a decline in quality of life … that drove important prosperity from [out of] the community" (Marsh 1994).

We can more deeply understand alienating discourse in the 1994 book *Santa Ana: An Illustrated History* (Marsh 1994), published in cooperation with the Santa Ana Historical Preservation Society (SAHPS). The book includes a chapter entitled "The Renaissance," which devotes most of its pages to Cribb's roughly 2,500-word essay, "How the Arts are Helping to Save Santa Ana." As he succinctly invokes a few city case examples, none of which resemble Santa Ana, he proclaims that his vision was an obvious choice for Santa Ana because the arts are a silver bullet solution to improve a host of urban quality of life issues. While he did not discuss how this would presumably happen, he did provide rationality and vision for the arts in downtown:

> The Santa Ana Council of Arts and Culture (SACAC) formed four years ago to generate for our city a cultural profile, attract human and material resources to the community, and focus attention on a future here in which arts and culture become increasingly beneficial to businesses, education, and residential components. Nationally, overwhelmingly evidence has shown that arts and cultural development arrests the decline in urban environments and builds bridges between the disparities of income, ethnic backgrounds, education, and life experience, all while creating a new atmosphere for growth and prosperity. (ibid., p. 162)

He went on to ask what may attract people to live and visit places. *What is it that draws people to Laguna Beach, parts of Los Angeles, San Francisco? Is it not the cultural districts with unique restaurants and shopping opportunities?* To borrow from Holston (1989), Cribb's imagined world resembles a transformation from an unwanted present to glitz and wealth.

Agustin Gurza, who reported on Santa Ana and the Latina/o community for the *Orange County Register* during the 1990s, took a more blatant approach to erasing, reimaging, and disparaging. Gurza wrote in 2009 in *Orange Coast*, a regional magazine targeting middle-income to upper-income County residents, that the Artists Village was much needed, and in the process noted:

> Santa Ana now stands for something, and it's not gangs or illegal immigrants. It stands for art that's edgy and experimental, informal and anti-establishment, as the artists themselves define it. (Gurza 2009)

Reimagining also occurred through the city's rebranding campaign during the RP period in the mid-2000s. For example, Lejano and González (2016) show one rebranding effort as a city marketing video, presumably for developers, investors, and businesses, which overwhelmingly depicts a middle-class and upper middle-class image, one showing that the city is representative of housing, businesses, and other amenities for these populations. For instance, the video shows a mother in her convertible with the top down and her son in the passenger seat driving through Santa Ana's most well-known middle-class neighborhood, Floral Park. Floral Park is home to more than 600 vintage homes, mostly built between the 1920s and 1950s, including over a hundred homes on the Santa Ana Register of Historic Properties (Floral Park Neighborhood Association 2016). The community boasts the city's highest median sales price, approximately $900,000 at the height of the property boom in 2007, while Santa Ana as a whole had a $585,000 median sales price housing in the same period. The story continues with this narrative, showing the city's BMW car dealership and fine dining and nightlife scenes, including the Artists Village. The marketing video reimagines the racial/ethnic composition of the city—overwhelmingly, the people in the video are non-Latina/o. The demographic of Floral Park itself is predominantly White. The images in the video are a far distance from the urban community depicted by the Nelson A. Rockefeller Institute of Government who in 2004 ranked Santa Ana first in its national Urban Hardship Index (Montiel, Nathan, and Wright 2004).

Another major rebranding activity by the city was changing the city slogan from "Education First" to "A Place for Art" and "Downtown Orange County." The most recognizable example of this is the restored and repainted water tower adjacent to the Logan neighborhood and a major freeway, reading "Downtown Orange County." Clearly, the city was invested in rebranding Santa Ana through the arts and by self-elevating the downtown district as the "hot spot" in the entire county.

The city's largest scaled plan for erasing and reimagining is through the RP, which Lejano and I detailed elsewhere (2009). The redevelopment plan engages in erasing and reimagining in a number of ways, of which I discuss some here. For instance, the plan's images and text show and mention almost nothing explicit about the Latina/o demographic of Santa Ana, particularly downtown. More precisely, the plan's section on the downtown neighborhoods devotes two pages to each barrio and lacks a description of their socioeconomic profile. It also fails to mention barrio history related to community development, or to note how residents may be accounted for in planning considerations. The plan also largely avoids any mention of *La Cuatro*'s social and economic profile and history.

There is reference to the word "Hispanic" once, and this reference is to argue that downtown's economically thriving cultural district success is in reality a deceptive assessment because it is overwhelmingly geared towards the city's majority demographic and missing an opportunity to improve. Here as well the plan omits mention of the word *La Cuatro*, erasing the area's well-established identity. For example, the introductory chapter states "Downtown Santa Ana is a thriving commercial district … and offers a broad range of goods and services focused primarily towards the Hispanic community" (Moule and Polyzoides Architects and Urbanists 2007). Shortly thereafter, the plan characterizes this economically thriving cultural district problematic because the area is "not meeting its potential" (ibid., 1:2). This comes with the mention that downtown merchants report strong sales.

The needs the plan talks about are primarily those related to new or potential entrants into the plan area, as identified by their ability to pay market rates, and to integrate to the area the wealthier and majority white county region in which Santa Ana is embedded. The plan stipulates:

> Santa Ana has recently become a proven location for the development of new market rate lofts and live work type of dwellings. These new homes are appealing to a wide range of buyers including young professionals, families, and empty nesters. These new residents appreciate Santa Ana's urban character, diversity and convenient lifestyle. This new residential trend is likely to continue and will expand the downtown's commercial market. Many of these new home owners have expressed a desire for more variety in retailers and restaurants. (Moule and Polyzoides Architects and Urbanists 2007)

The plan prioritizes and normalizes the need to attract new residents from the county. We see this normalization when the plan characterizes downtown Santa Ana as a "unique opportunity within the greater Orange County region that is perceived as not meeting its potential" (ibid., 1:2). While the plan does not entertain who or what groups have this perception, these residents would, by implication, improve perceptions that Orange County is some type of up and coming region. New visitors and residents to Santa Ana, then, is a key to improve or replace some general non-favorable perceptions about Orange County.

The plan also erases *La Cuatro*'s cultural business element in favor of cultural elements that are not represented in the zone (González and Lejano 2009). The needs of a representative base of *La Cuatro* business owners are missing from the plan's text and imagery, as land use pictures do not specify how they reflect this major sector. As an example, the chapter does not provide a representative sample of businesses or business signage in the area. Instead, the plan discusses possible new business signage, and the pictures illustrating acceptable signage show the following: "The Talking Room: Classic Café Dinning" and "Fine Oriental Carpets and Kelims" (Moule and Polyzoides Architects and Urbanists 2007). These examples provide cultural cues that do not match the area. Instead, consider a sampling of signs that one would have seen in the downtown during this time period: "Ritmo Latino," "Teresa's Jewelers," and "Minas Bridal."

The plan's form-based codes (FBCs) also construct a reimagined community through physical design (González and Lejano 2009). The RP adopts a "science of form" agenda that socially constructs the urban center in an altogether reimagined community (ibid.). The plan's FBCs are a type of science and measurement to objectify the target population and embed disciplinary mechanisms into every fact of life (Foucault 1977, 1991). As an example, the plan assumes specifications to which the present community does not conform. This includes the existing density, in persons per household, being significantly larger than that assumed in the plan. "Intensity" is defined as a range from "the most urban types of development and land use to the least urban types" (Moule and Polyzoides Architects and Urbanists 2007). This intensity is operationalized in residential areas by measuring dwelling units per acre. For the two neighborhoods, the RP poses two conflicts in this respect. One, overly intense land uses in residential areas are incompatible with industrial and residential uses. In some situations, residential intensity is too high; in other cases, industrial intensity is too high. The most immediate concern is non-conformity with intensity criteria. The RP stipulates that residential/industrial and institutional zones should have the lowest intensities. Moreover, the plan disallows multiple families in two building types which are common in the city and area: "house" and "use/work."

We can see how intensity is problematic in light of Santa Ana's demographics. Santa Ana has one of the highest population densities in the country. At the same time, the city's building stock consists of mainly single-family homes. The discrepancy is resolved when we understand that many of the single-family homes house multiple and intergenerational families (The Planning Center 2008). This is one strategy for coping with high real estate prices. Also relevant is the high proportion of undocumented families in Santa Ana who suffer from any increased attention to population density. In this light, the RP's construction of intensity as problematic can be threatening to many residents.

Just like Swanson's notion of "mixing," language in the RP around diversifying Santa Ana was perceived as threatening to many local activists. In this context, we can appreciate local writer Arellano's view that:

> For years, Santa Ana officials have dreamed out loud about "diversifying" and "revitalizing" the city's downtown—code for driving out Mexican businesses in favor of chains and hipsters. After years of planning, city leaders now have a proposal: the Renaissance Specific Plan (RSP), which a city PowerPoint presentation states is "a planning tool to guide development" and "translates the community vision into goals, policy initiatives, regulations, and standards." In other words a way to drive out Mexicans. (Arellano 2008a)

He goes on to comment on the architectural templates found in the RP and writes: "With the exception of the last [template] one (which is the preferred style for overpriced lofts nationwide), each had its heyday before World War II—which, in SanTana history, was a time when everything was wonderful and the darkies and brownies couldn't enter certain restaurants, had to live in certain neighborhoods, and to go to the balcony to see movies" (Arellano 2010).

The broader real estate community has also contributed to the spatially alienating discourse. For example, the Urban Land Institute (ULI) presented a redevelopment vision before a full audience at the Yost Theater in November 2010. I attended this presentation. The presentation, "Diagnosing Place: The Hidden Epidemic," was designed to convince the public in attendance, largely dressed in corporate and business casual attire, that Orange County lacks a central place identity—a downtown—and that they should join a movement to create one. The presentation included many images of downtown Santa Ana, though these were mostly digitized and made generic to avoid targeting *La Cuatro*. The presentation included images of a trolley zipping through neighborhoods and downtowns. In the event's Question and Answer session, the first question from the audience asked how the redevelopment ideas and placemaking strategies, which were heavily inspired by new urbanist and creative city principles, could work in and around *La Cuatro*, particularly given past and ongoing controversy about gentrification. Urban Land Institute representatives understood this was a hot topic locally. With a brief pause and looking towards the floor, and a bit uncomfortably, the representative succinctly told the audience members to "embrace" the local context and immediately moved on to other questions.

Spatial Alienation and Urban Politics

Spatially alienating mechanisms and ensuing practices can come to fruition and maintain relevancy via city–developer improprieties, despite protest. Much of the visible changes to and marketing for the recently branded East End have, in part, come from very generous subsidies from the city to the East End developer and this developer's leadership in the local Community Management District that was approved by the city. For example, in 2008 the city awarded $765,000 of the $1,250,000 "Fourth Street Façade Program" to Fiesta Marketplace Partners (Orange County Grand Jury 2012). One of the original partners, Irving Chase of S and A Properties, had at this time gained 98 percent majority interests in the Fiesta Marketplace Partners (ibid.). Around this time, S and A Properties began an aggressive rebranding campaign of Fiesta Marketplace to the East End. This included actively soliciting new types of small businesses and retailers that belong or market to the cultural genotypes typically associated with the creative class. Irv Chase also told his tenants, namely those who had been operating during the Fiesta Marketplace entity, that he would not renew their leases unless they expanded their merchandise to include items that could appeal to groups other than Latina/o immigrants (Elmahrek 2011b). Ascribing changing types of businesses to this practice, Elmahrek describes the changes in this manner:

> Santa Ana's Fourth Street in the heart of the city's downtown is known for its medley of Latino-owned shops selling goods like bridal gowns, ranchero style boots and gold jewelry along storefront displays. But once you get to the blocks between Bush and French streets [East End], the beginning of change can be seen. Style World, a women's clothing shop, has a row of manikins in its display that flaunt higher-end lines of clothing. The attempt, the store manager said, is to look like "Forever 21," a corporate chain that caters to young women. Style World isn't alone. A barber shop that looks like it was plucked form a trendy urban district has also popped up. And so has a tattoo and piercing supplies shop along with a new coffee shop. The carousel that harkened memories of childhood for many Santa Ana residents is gone and nearby building facades are starting to take on the look of a South Orange County shopping plaza. (Elmahrek 2011b, p. 1)

Some in the broader community suspected that improprieties had driven those responsible for the funding and decision-making behind these changes to the East End. The community suspected the formation of the downtown Community

Management District, also established in 2008 and known locally as the Property Based Improvement District (PBID), had happened illegally. Borrowing heavily from urban policy strategies that are designed to attract the creative class (Borrup 2006), the PBID occupied sixty-six blocks of downtown including the East End. The PBID was the main economic engine and decision-making body behind creative city programming and marketing efforts for the East End. The district designation allowed a few property owners to establish an assessment/property tax on all property owners within the district that generates funds to support services over and above those provided by government, including maintenance, security, marketing, promotions, and general economic development. Downtown Inc., a private nonprofit, was formed to oversee the PBID and redevelopment (Orange County Grand Jury 2012). Irving Chase and his son, representing the Partners, had from the start at different times held the position of President for Downtown Inc.

Almost immediately, some businesspeople within the district and activist allies argued that Downtown Inc. significantly steered marketing funds to attract creative class–type commercial and cultural activities and target consumers who would be drawn to such activities, especially to the East End. Very little resources were directed towards *La Cuatro* and its target demographic, these groups added. Some of the protesting business owners displayed storefront signs proclaiming "Stop PBID" (see Figure 3.5).

In response to protest against the PBID and accusations that he was driving out Latina/o immigrant businesses and consumers with changes to Fiesta Marketplace, Chase stated during a City Council meeting that his business decisions were

Figure 3.5 Storefront posters in protest of the Property Based Improvement District (PBID).

Source: Author's personal collection.

strictly economics, stating, "I'm really sick and tired of the race issue being played here over and over again.... I want to remind the council that I'm the one who built Fiesta Marketplace" (Elmahrek 2011d, p. 1). He stressed that he had not made a "penny" on the properties and personally invested more than 5 million in remodeling Fiesta Marketplace into East End (Elmahrek 2011d, Pimentel 2014). Chase had divorced himself from his Fiesta Partners and original visionaries and any subsidy support for Fiesta Marketplace and the East End. Instead, he portrayed himself as a one-time downtown rescuer for the Latina/o downtown market, having made not one cent of profit in the process and now looking to exercise his vision for personal and downtown economic success.

A group of businesses did not sympathize or empathize with the investor and took legal action against the PBID. The Orange County Grand Jury found that accusations against S and A Properties had legitimate grounds. In 2012, the Orange County Grand Jury, which was composed of nineteen members who lived in the county's five supervisorial districts, concluded in an official report that the city of Santa Ana "appears to be in violation of California State law in the formation of this Improvement District" (Orange County Grand Jury 2012). The grand jury found that "monies collected from the improvement district appear to have only benefited a few and have not resulted in a direct benefit to the assessed property as required by California law" and that "an appearance of impropriety exists in the relationship between the developer and the City of Santa Ana." The report also pointed out that the vast majority of dollars from the Fourth Street Facade Program went to the same developers "who were the primary interests in changing the culture of the area" (ibid., p. 2). The report also insinuated that network politics added to these decisions and pointed out that the same developers had "extensive connections to Downtown Inc. which the city chose to manage the proceeds from this special assessment, such that they had at various times held positions as the officers and directors" (ibid.). The findings also reported "irregularities" in the establishment of the PBID and council members repeatedly ignored a petition to disestablish the PBID by owners of property within the district (ibid., p. 10).

A Street-Level View of Downtown Change and Continuity

All the redevelopment controversy and business changes in the downtown area prompted me to understand the extent to which businesses were changing and where. By now, there were plenty of qualitative descriptions in the media about the changes that were transpiring downtown. This type of qualitative description is also quite common in the commercial gentrification literature. I wanted to compliment anecdotal descriptions with a systematic and quantitative street-level view of the number and type of businesses one can actually find across *La Cuatro*, the Artists Village, and the East End.

"In my 2011 survey of downtown businesses[1]," I found 464 businesses that make up the study area (González and Guadiana 2013). While the Artists Village, East End, and La Cuatro have similar percentage of businesses in the food, dining, and liquor category (11%, 12%, and 5%, respectively), they have key dissimilarities, highlighting patterns of redevelopment targeting a reimagined community.

The Artists Village, for example, provides relatively distinctive subtypes of business options relative to *La Cuatro*. The majority of the food businesses in the Artists Village are bistros, cafes, late-hour bars, and general restaurants. Performing arts companies, entertainment centers, and galleries and studios are also common, consisting of eight percent of business types in the zone. These businesses collectively cater to the creative class and middle class. We can understand this alienation further with a specific example, the Cooper Door bar. This bar opened in the late 2000s in a location previously held by dive bar Broadway Billiards. Cooper Door is a modernly refurbished bar with extensive craft beer poured in rustic mugs, and a long pine communal table. The prior billiards dive had catered to locals, with pool and carambola tables. Its television routinely showed soccer or Spanish-language programming. The bar sold affordable domestic and Mexican beers, Cup Noodles, popcorn popped to order in a microwave, Tapatio hot sauce, and lime, and had a jukebox heavy in Spanish-language music. Art galleries and shops are other examples of businesses that cater to a reimagined population. While a few of the art galleries and shops in the Village have been Latina/o-owned at times, selling medium- to high-priced Latina/o multigenerational culture-themed artwork, suggesting a match with *La Cuatro*, overall these business cater to the creative class consumers in the area who are not Latina/o.

The East End, while not having the same overrepresentation of businesses catering to an imagined community as the Artists Village, is increasingly becoming like the Village. The majority of businesses in the East End specializing in clothing and accessories (15%), followed by businesses selling food. Both types of businesses cater to the working-class immigrant demographic, though this business profile is transitioning. For instance, most of these businesses pre-date the East End redevelopment and rebranding activity under Fiesta Marketplace and have Spanish-language names, such as Fiesta Sportswear, El Paso Shoes, La Pizza Loca, El Rancho Restaurant, and Taqueria Guadalajara. Some new businesses, however, have opened more recently as part of the East End strategy and cater to the creative class. One new dining establishment, the Playground, replaced a Mexican seafood restaurant, Mariscos Tampico, that was lower priced. And, in place of Moya's Panaderia, plans were being made for a trendy "Dough Exchange" (Goei 2013).

Other practices that have helped displace the Latina/o and working-class and immigrant community were changes to urban design during the Fiesta Marketplace to East End change. The kiosk, carousel, and a seating/socializing area were removed and replaced with an open promenade. As I described in chapter two, these amenities were intended to attract immigrant families. Another change was the replacement of the colorful pastel tones of the commercial buildings that used to represent a symbol of Mexican architectural design with neutral tones. Figure 3.6 and 3.7 show Fiesta Marketplace in 2002 and 2008 respectively, looking north, and Figure 3.8 shows the newly redesigned East End in 2011, also looking north.

While *La Cuatro* mainly contains real estate and other investment-type establishments (16%), such as "CM Properties" or "Sycamore Partnership," most of which are not open for walk-in business, collectively more common business types at just over 50% and open for business as well are administrative, insurance,

Figure 3.6 Fiesta Marketplace looking north in 2002.

Source: Courtesy of the Santa Ana History Room/Santa Ana Public Library.

Figure 3.7 Fiesta Marketplace kiosk looking north in 2008.

Source: Author's personal collection.

Figure 3.8 East End looking north in 2011.
Source: Author's personal collection.

and financial services, followed by independent vendors; bridal and quinceañera stores; clothing and accessories; and beauty salons, barber shops, and supplies, many which have Spanish-language names and collectively target the working-class immigrant population. Some examples of these businesses include Aztlan Insurance, La Zapateria Mexico, and Beauty Salon Nuevo Guadalajara.

La Cuatro's use of public space is also unique to the other two zones in three ways. For example, regulated pushcart vending on select corners help transform public space into public mini-markets and gathering locations. Vendors sell anything from *churros* to *elotes* (corn on the cob), *bionicos* (fruit salad), hot dogs, chips, and soft drinks (Figure 3.9). Marketing methods are also unique to the area. Some storeowners extend their street marketing by having an employee stand outside in front of the business and create awareness of the business in hopes of attracting customers. Employees, often women, hand out business cards or flyers, promoting a product or service, encouraging people to enter the establishment, and inviting conversation (Figure 3.10). For example, in my 2011 field observations, an employee from Jacobo Foto Y Video said "Pásele. Que buscas? Si quieres ver precios, te dan un precio aqui arriba" [Come on in. What are you looking for? You can get prices upstairs if you wish.]. Another example is an employee from Jessica's Bridal, who said "Buscaba algo en especial? Pásele por favor" [Looking for something in particular? Please come on in.]. During the time of the survey, I found 15 businesses generally spread out along *La Cuatro* using this type of marketing strategy on a Saturday afternoon and this amount of businesses using this type of street marketing is more or less typical on weekends. Regulated pushcart vendors and street marketing are generally not present in The East End and the Artist Village.

Figure 3.9 Street vendor on *La Cuatro's* Fourth St. and Main St., the area's busiest intersection.

Source: Author's personal collection.

Figure 3.10 A business employee hands out flyers.

Source: Author's personal collection.

As a whole, *La Cuatro* still maintains its identity, heavily targeting the working-class immigrant population; however, in the past decade the zone has seen change. There has been growth of new businesses like those in the Artists Village and those opening in the East End, but mainly around the edges contiguous with those zones. The threat of alienating business types to *La Cuatro* is, as I have chronicled, a real lived experience to business owners in the area and anti-gentrification activists, one that could begin to encroach inward toward the thriving heart of the downtown.

Note

1 We developed typologies for business types and describe the most common business categories in the area. We report on the six most common business categories among the three zones, accounting for 49 percent of all businesses.

References

Arellano, Gustavo. 2007a. "Santa Ana's small-minded self-hatred." *Los Angeles Times*, August 2, 2007. Op-Ed. www.latimes.com/la-oe-arellano2aug02-story.html. Access date August 10, 2007.

Arellano, Gustavo. 2007b. "The renaissance blog in Santa Ana's redevelopment plan." *Naval Gazing: The World Famous OC Weekly Staff Blog*. December 28, 2007. http://blogs.ocweekly.com/navelgazing/2007/12/the_renaissance_blob_in_santa.php. Access date December 29, 2008.

Arellano, Gustavo. 2008a. "The renaissance blob." *OC Weekly*. January 10, 2008. www.ocweekly.com/news/the-renaissance-blob-6398785. Access date January 15, 2008.

Arellano, Gustavo. 2008b. "The renaissance reconquista will be televised." *Naval Gazing: The World Famous OC Weekly Staff Blog*. August 26, 2008. http://blogs.ocweekly.com/navelgazing/2008/08/the_renaissance_reconquista_wi.php. Access date August 29, 2008.

Arellano, Gustavo. 2008c. "Why Santa Ana natives fear the hipsters." *Naval Gazing: The World Famous OC Weekly Staff Blog*. January 17, 2008. www.ocweekly.com/news/why-santa-ana-natives-fear-the-hipsters-6483120.

Arellano, Gustavo. 2010. "Santa Ana's re-renaissance plan wants to Marty McFly us back to the past!" *OC Weekly*. March 24, 2010. www.ocweekly.com/news/santa-anas-re-renaissance-plan-wants-to-marty-mcfly-us-back-to-the-past-6451789. Access date April 2, 2010.

Becerra, Hector. 2008. "Latino yes, but new tastes." *Los Angeles Times*. May 28, 2008. http://articles.latimes.com/2008/may/28/local/me-amigostores28. Access date May 29, 2008.

Borrup, Tom. 2006. *The creative community builder's handbook*. Saint Paul, MN: Fieldstone Alliance.

Carcamo, Cindy. 2015. "Santa Ana leaders encourage young Latino activists to speak, by ballot." January 5, 2015. www.latimes.com/local/orangecounty/la-me-ff-santa-ana-latino-divide-20150105-story.html. Access date January 7, 2015.

Chang, Richard. 2006. "Open house in Santa Ana worth checking out." *The Orange County Register*, November 12, p. 15, Arts and Entertainment. www.occca.org/exhibition-images/2006/11-New-Media/New Media review.jpg.

City of Santa Ana. 2007. "Santa Ana Renaissance Specific Plan: frequently asked questions-redevelopment issues." www.ci.santaana.ca.us/pba/planning/documents/SARSP-FAQs_Redevelopment.pdf. Access date February 12, 2008.

City of Santa Ana. 2008. City of Santa Ana: creating community together.

City of Santa Ana. 2009. City of Santa Ana General Plan: Housing Element 2006–2014. www.santa-ana.org/housingelement/documents/Housing.pdf.

Community based coalition addressing the Renaissance Plan. 2007. Meeting minutes.

Delson, Jennifer. 2006. "Santa Ana City Council is all-Latino." *Los Angeles Times*. November 9, 2006. http://articles.latimes.com/2006/nov/09/local/me-santaana9. Access date November 12, 2008.

Delson, Jennifer. 2008a. "Re-map excludes officials' holdings." *Los Angeles Times*, January 20, 2008. http://articles.latimes.com/2008/jan/20/local/me-pulido20. Access date January 22, 2008.

Delson, Jennifer. 2008b. "Plan seeks to make Santa Ana the heart of Orange County." *Los Angeles Times*. January 20, 2008. http://articles.latimes.com/2008/jan/20/local/me-downtown20.

Diaz, David R. 2012. "Barrios and planning ideology: the failure of suburbia and the dialects of new urbanism." In *Latino urbanism: the politics of planning, policy, and redevelopment*, edited by David R. Diaz and Rodolfo D. Torres, 21–46. New York: New York University Press.

Downtown Property Owners and Merchants Working Group. 2007. Renaissance Plan–Request to Collaborate in Its Design. May 24, 2007. Letter submitted to City of Santa Ana City Council.

Dubin, Zan. 1997. "Santa Ana Artists Village edging closer to reality." *Los Angeles Times*. October 31, 1997. http://articles.latimes.com/1997/oct/31/news/mn-48695. Access date May 2, 2013.

Elmahrek, Adam. 2012. "Grand jury: Downtown Santa Ana's promotions tax violates state law." Voice of OC. June 29, 2012. http://voiceofoc.org/2012/06/grand-jury-downtown-santa-anas-promotions-tax-violates-state-law/. Fourth Street property owner puts heat on merchants

Elmahrek, Adam. 2011a. "Something's happening in Santa Ana's downtown." *Voice of OC*. May 31, 2011. http://voiceofoc.org/2011/05/somethings-happening-in-santa-anas-downtown/. Access date May 31, 2011.

Elmahrek, Adam. 2011b. "Fourth Street property owner puts heat on merchants." *Voice of OC*. April 14, 2011. http://voiceofoc.org/2011/04/fourth-street-property-owner-puts-heat-on-merchants/. Access date April 14, 2011.

Elmahrek, Adam. 2011c. "Omission in Santa Ana downtown tax law riles property owners." Voice of OC. February 12, 2011. http://voiceofoc.org/2011/02/omission-in-santa-ana-downtown-tax-law-riles-property-owners/.

Elmahrek, Adam. 2011d. "Santa Ana councilwoman steps into gentrification battle." *Voice of OC*. July 10, 2012. http://voiceofoc.org/2011/07/santa-ana-councilwoman-steps-into-gentrification-battle/. Access date July 19, 2011.

Floral Park Neighborhood Association. 2016. "Floral Park Neighborhood Association." http://floral-park.com/. Accessed March 23, 2016.

Foucault, Michel. 1977. *Discipline and punish: the birth of the prision*. New York: Pantheon.

Foucault, Michel. 1991. "Governmentality." In *The Foucault effect: studies in governmentality*, edited by Graham Burchell, Colin Gordon, and Peter Miller. Chicago: University of Chicago Press.

Galvin, Andrew. 2011. "Ethnic cleansing invoked by councilwoman." *Orange County Register*. August 25, 2011. http://santaana.ocregister.com/2011/08/25/hitler-ethnic-cleansing-invoked-by-councilwoman/. Accessed August 25, 2011.

Garlington, Phil. 2000. "Creative differences in Artists Village cities: Santa Ana district isn't shaping up the way some would like. But not all agree on solutions, or even that's there's a problem." *Orange County Register*. Feburary 6, 2000. Access date June 15, 2012.

Goei, Edwin. 2013. "Playground working on a standalone bakery next door, taking over former Mexican bakery booted out by gentrification." *OC Weekly*, October 9, 2013.

www.ocweekly.com/restaurants/playground-working-on-a-standalone-bakery-next-door-taking-over-former-mexican-bakery-booted-out-by-gentrification-6612567. Access date November 2, 2013.

González, Erualdo R., and Lorena Guadiana. 2013. "Culture oriented downtown revitalization or creative gentrification?" In *Companion to Urban Regeneration*, edited by Leary M. E. and J. McCarthy. Oxon, UK: Routledge.

González, Erualdo R., and Raul P. Lejano. 2009. "New urbanism and the barrio." *Environmental and Planning A* 41 (12):2946–63.

Greenhut, Steven. 2007. "A scheme bad to the core." *Orange County Register*, December 2, 2007. www.ocregister.com/ocregister/opinion/columns/article_1947873.php. Access date December 3, 2008.

Greenhut, Steven. 2008. "One small step for liberty: Santa Ana renaissance plan may be halted, which is a small step ahead for local freedom." February 24, 2008. *Orange County Register*. www.ocregister.com/ocregister/opinion/columns/article_1986551.php. Accessed February 29, 2008.

Gurza, Agustin. 1995. "Setting lofty goals/cities: Santa Ana hopes to create an Artists Village to return downtown buildings to their former glory." *Orange County Register*. June 1, 1995. Accessed July 12, 2013.

Gurza, Agustin. 2009. "The resurrection of Santa Ana." *Orange Coast,* August, pp. 82-8.

Holston, James. 1989. *The modernist city: an anthropological critique of Brasilia*. Chicago: University of Chicago Press.

Irving, Doug. 2010. "Campaign money raises questions over Santa Ana vote." *Orange County Register*, September 4, 2010. http://m.ocregister.com/news/vote-266466-city-council.html.

Irving, Doug. 2011. "Groups seek grand jury probe of Santa Ana." *Orange County Register*, January 5, 2011. www.ocregister.com/news/coalition-282858-grand-jury.html. Accessed January 5, 2011.

Lejano, Raul P, and Erualdo R. González. 2016. "Sorting through differences: the problem of planning as reimagination". *Journal of Planning Education and Research* 0739456X16634167, first published on March 23, 2016 as doi:10.1177/0739456X16634167.

Londoño, Johana, and Erualdo R. González. 2015. "The changing politics of Latino consumption: debates related to downtown Santa Ana's new urbanist and creative city redevelopment." In *Race and retail: consumption across the color line*, edited by Mia Bay and Ann Fabian. New Brunswick, NJ: Rutgers University Press.

Mantle, Larry, and Derrick Turac. 2011. Santa Ana's gentrification sparks debate, March 29, 2011. 89.3 KPCC. www.scpr.org/news/2011/03/29/25430/santa-anas-gentrification-wars-discussed-airtalk/.

Marsh, Diann. 1994. *Santa Ana … an illustrated history*. 2nd ed. Untied States of America: Heritage Publishing Company.

Medina, Jennifer. 2011. "New faces and contentious revival." *The New York Times*, October 30, 2011. www.nytimes.com/2011/10/31/us/in-santa-ana-new-faces-and-a-contentious-revival.htmlpagewanted=all&_r=0. Accessed October 30, 2011.

Mena, Jennifer. 2002. "Santa Ana hopes to get a lift from lofts." *Los Angeles Times*, September 14, 2002.

Mills, Sean. 1999. "Coffee shop crack leaves bitter taste." *Los Angeles Times*

Mills, Sean. 2004. "Ethnic cleansing in Santa Ana?" *The Orange County Register*.

Montiel, Lisa M, Richard P Nathan, and David J Wright. 2004. *An update on urban hardship*. Albany, NY: The Rockefeller Institute of Government.

Moule and Polyzoides Architects and Urbanists. 2007. Santa Ana Reinassance Specific Plan: Draft.

Orange County Center for Contemporary Art (OCCCA). 2016. "Orange County Center for Contemporary Art (OCCCA)." http://occca.org/. Accessed February 4.

Orange County Grand Jury. 2012. Santa Ana's Property Based Improvement District. www.ocgrandjury.org/pdfs/cityofsantaanareport.pdf.

Parsons, Dana. 2008. "Will Santa Ana's opposition be heared above the din?" *The Los Angeles Times* June 20, 2008. www.latimes.com/news/local/orange/la-me-parsons20-2008jun20,0,7806980.column.

Peck, Jamie. 2005. "Struggling with the creative class." *International Journal of Urban and Regional Research* 29 (4):740–70.

Pimentel, Joseph. 2014. "Trendy crowding out traditional in Santa Ana." *Orange County Register*. September 21, 2011.

Roberts, Jerry. 1998. "Art for city's sake.": The Santa Ana Artists Vilage is transforming a blighted downtown into a vibrant creative center. *Orange Coast*, December, pp. 56–63.

Rojas, Leslie B. 2011. "Will Latino-led redevelopment in Santa Ana ultimately rob it of its Latinoness?" 89.3 KPCC Radio. Accessed April 20, 2011. http://multiamerican.scpr.org/2011/03/will-latino-led-redevelopment-in-santa-ana-ultimately-rob-it-of-its-latino-ness/.

Segal, Dave. 2007. "Three the hard way." *OC Weekly*. September 13, 2007. www.ocweekly.com/music/three-the-hard-way-6398978. Accessed March 2, 2010.

Taxin, Amy. 2006. "Santa Ana looks for crossover business." *Orange County Register*. September 6, 2006. www.ocregister.com/articles/downtown-58172-city-santa.html.

Thakur, S. 2008. Physical and economic analysis of Santa Ana Renaissance Plan. University of California, Irvine.

The Kennedy Commission. 2008. City of Santa Ana land held for future development: Community Redevelopment Agency annual report, fiscal year 2006–2007. Irvine, CA.

The Planning Center. 2008. City of Santa Ana General Plan: Draft Housing Element. www.ci.santa-ana.ca.us/news/Preliminary draft Housing element-0908.asp.

Thomas, June. 2008. "The minority-race planner in the quest for a just city." *Planning Theory* 7 (3):227–47.

U.S. Census Bureau. 2000. State and County Quickfacts: Santa Ana, CA. Washington, D.C.: Government Printing Office.

U.S. Census Bureau. 2010. "The Hispanic Population: 2010." U.S. Department of Commerce. www.census.gov/prod/cen2010/briefs/c2010br-04.pdf.

U.S. Census Bureau. 2011. American Fact Finder, Selected Social Social Characteristics in the United States, 2011 American Community Survey 1-Year Estimates. U.S. Department of Commerce. Santa Ana, CA.

4 The Grassroots Rises, 2000s

In late 2007, a group of political and cultural empowerment Latina/o organizations, local activists, and some local residents began meeting regularly at La Chiquita Restaurant, a locally owned Mexican restaurant, and Chepa Park community center in the Logan Barrio, a neighborhood in the proposed Renaissance Specific Plan (RP). I was part of those meetings and while none of us were certain initially how we would collaborate and under which specific objectives, there was a common suspicion that the RP, which was slated for city vote within a few months from our first gathering, was the city's latest, and most sophisticated, scheme to reimagine and alienate the downtown district. Everyone in the meeting felt an urgency to study the plan and its potential ramifications for the downtown district, especially for the two neighborhoods included in the plan. The group, which subsequently formed into the Santa Ana Collaborative for Responsible Development (SACReD), understood that threats of this nature had been all too common in the downtown's past. Unclear to the group, however, was exactly how this contemporary threat would manifest in practice and how the group would respond.

This chapter details the role of the grassroots in articulating and responding to this threat. Two questions guide this chapter. How do community-based approaches address the needs and hopes of lower-income Latina/o urban areas undergoing Transit-Oriented Development (TOD) and New Urbanism (NU)? What are the challenges, contradictions, and limitations in practicing community-based participation?

Public–Private Collaborative Planning for the Station District

In 2009, the city hired Related California/Griffin Realty Corporation (hereinafter "developer") as the master developer for a Station District (SD) development project and formalized a Predevelopment Agreement for planning and development purposes. According to city records, the developer was well qualified for developing residential for sale, commercial, retail, mixed-use projects, and rentals in all ranges of income, securing financing, and obtaining community participation (City of Santa Ana 2009). The city mandated that the developer take charge of neighborhood outreach and participation, and coordinate these efforts with the city's Neighborhood Improvement Division (ibid.).

The SD concept plan was partly prepared with input from about 200 key stakeholders in the area who participated in community outreach meetings (Related, Griffin Realty Corporation, and WHArchitects 2010). The meetings were conducted over two months with the Lacy Neighborhood Group, Logan Neighborhood Group, French Park Neighborhood Group, Santiago Street Lofts Homeowners, St. Joseph's Church, and Station District Business Owners. The stakeholders also participated in a bus tour to see examples of other projects that the developer had completed. City staff supplemented these participation activities and held an additional 20 community meetings and interviews, and "also worked diligently on an ongoing basis with SACReD … to obtain input on the proposed development and planning effort" (City of Santa Ana 2010). The developers also held "public review/comment and revisions to development plan" for city staff, City Council, and stakeholder input prior to finalizing the plan (City of Santa Ana 2009, Related, Griffin Realty Corporation, and WHArchitects 2010).

On January 10, 2012, the City of Santa Ana Redevelopment Agency and the developer held a groundbreaking ceremony for the SD. Mayor Pulido noted at the groundbreaking that the SD is:

> [An] outgrowth of a comprehensive and collaborative approach to planning and development. Careful planning, selection of the right development team, and the involvement of the community has created a remarkable housing opportunity, one that adds to the vitality of the neighborhood and city as a whole. (City of Santa Ana 2012)

Elected officials also acknowledged SACReD's presence in the audience.

The Birth of SACReD

SACReD attempted to halt the SD because it was an outgrowth of the RP. SACReD argued that both projects are one in the same, one member noting "I call it Renaissance Two" (Irving 2010b). The group feared that the SD would pave the way for more redevelopment similar to new lofts intruding on the Lacy and Logan barrio borders and businesses downtown, both designed for regional middle-class entrants. The group had an issue with this vision; these, it argued, were a mismatch for the dominant demographic and realities in the two neighborhoods. As an example, the majority or representative demographic from the Lacy neighborhood would not be able to afford market-rate lofts.

In September 2007, Chican@s Unidos de Orange County, a grassroots group, obtained collaboration from the nonprofit and all-volunteer-run El Centro Cultural de México, Barrio Logan activists, and others and convened meetings to study the Renaissance Plan and draft responses to the plan's upcoming Environmental Impact Report. The meetings were held in English with Spanish translation provided when needed. The organizing of residents and a few local business owners from *La Cuatro* also began at this time, though their overall attendance in coalition meetings was sporadic and minimal. Concurrently, efforts arose from nonprofit organizations that had ongoing ties with residents in the plan area by way of their own community organizing grant mandates. Community-based nonprofits including Latino Health Access, Orange County Congregation Community Organization (formerly the Santa Ana Neighborhood Organization, SANO), the Kennedy Commission (a housing nonprofit), and the Public Law Center were similarly driven to create a voice for the grassroots community and to initiate a plan for resistance to the city's redevelopment efforts. Within a month, all stakeholders formed a coalition to serve the interests of Lacy neighborhood residents. Most in the collaborative were familiar with each other, having worked together on other advocacy campaigns or cultural events in the city, some of this work dating to the late 1970s and 1980s. From the start, the stakeholders promoted grassroots leadership development and social justice.

Racial/ethnic and class identity politics drove the collaborative to explore a working relationship (González et al. 2012). The coalition understood that its work would be a deeply politicized process requiring a wide coalition of neighborhood and county stakeholders. Some of the coalition activists recognized that a Mexican/Latina/o and working-class community identity would be a critical and primary feature of their activism. Two Logan barrio activists involved in SACReD routinely made the collaborative aware of the need for collective action to preserve the right of the Mexican-Americans and Mexican immigrants to remain in the neighborhood. Logan barrio residents have engaged in rather routine protracted struggles for at least the last sixty years to preserve residential housing from repeated efforts to demolish housing and rezone the area for commercial use. More recently, Logan activists and a nearby neighborhood launched a lawsuit against neighboring French Park, which lobbied the city to install barricades to block incoming traffic to and from these neighborhoods (United States District Court, Central District of California, Southern Division 2004). Still other members traced their commitment to racial/ethnic and working-class politics to their direct involvement in SANO's Alinsky-style community organizing period, discussed in chapter two. The coalition also consisted of a younger generation of activists. Collectively, these members shared or empathized with local historical experiences of *barriozation*—community development and redevelopment oppression and exclusion common of low-income Mexican communities in the Southwest (Diaz 2005, Villa 2000). Such sentiments pushed a number of coalition members to create and wear pins with the slogan that invoked ethnic politics and the interconnectedness between Mexican neighborhoods and *La Cuatro*: "Stop Ethnic Cleansing! In Downtown Santa Ana." Clearly, there was general consensus from the start among SACReD members that they were committed to battle the city and developers and their activism would be largely driven by racial/ethnic and class identity politics and analysis.

Challenging the City's Plans for Redevelopment

The collaborative was determined to change the way the city conducts standard redevelopment practices, though the group was unclear early on what this change or changes would be and how exactly it would accomplish the objective.

Generally, some of the early thinking was about altering the city's and private developer's community participation routines.

SACReD believed that radical changes to the city's redevelopment status quo required bringing redevelopment practice under popular control, and the coalition could lead this charge (González et al. 2012). The idea was that popular control could infuse dignity, or respect, for the dominant area residents in the redevelopment process. Respect meant that local residents and activists could have meaningful involvement in redevelopment decisions over policies and resources and that the neighborhood's majority demographic would benefit equitably from such redevelopment. To reflect this perspective, the collaborative adopted the mission of *Unity is SACReD – Building a Dignified Community for All*. The group's goal emphasized responsible development, which it defined as the inclusion of residents, local community-based organizations, and businesses as equal partners and decision-makers in development negotiations. SACReD's goal was also about introducing a new social-policy contract in the form of new private–public participation approaches and a community benefits agreement (CBA) with the city. This was a historic grassroots and urban planning action in the city's storied and contested history with redevelopment.

SACReD accelerated activism when the city released the SD concept plan. The collaborative raced to create clear objectives and timelines. The coalition's main objective became to obtain benefits from the development that would address the needs of the local community, and would do so first by reforming private–public participation practices. Over the course of some months, members began to gather data and information within the Lacy neighborhood to inform the drafting of the collabortive's demands, thereby adding specificity to the group's stance for "responsible development." Over the course of about two years, SACReD designed and carried out a self-directed survey to understand resident attitudes towards and experiences with recent redevelopment and participation activities, held community forums and collected information during one-on-one organizing, and held debate and consensus among coalition members during meetings. As a whole, these processes informed the drafting of our community demands. But, SACReD also realized data collection would likely not translate into policy action without political activity. Around the same time, some SACReD members were meeting with selected city elected officials and high-ranking staff to express to them that SACReD was highly dissatisfied with the formal public–private participation process and to explore ways to involve SACReD in SD redevelopment decisions. SACReD learned during these meetings that the council, as a whole, believed that the SD project was necessary and could, with some modifications, address some community concerns that emerged from the RP. For example, one council member expressed gratitude to business owners and activists who organized against the Renaissance Plan, thanking them for "being part of the process" (Greenhut 2008). The City Manager, Dave Ream, requested models of community participation from the collaborative. Likewise, the collaborative was invited to participate in the SD design team and was hopeful it would be able to incorporate its demands into the conceptual plan. Subsequently, there were never serious discussions over the models and participation in the design team was restricted to two SACReD members previewing the conceptual plan. SACReD was adamant that this participation was tokenism and would not help the group work toward its objectives.

The Historic Social Contract the City Rejected

The city refused to consider SACReD's community participation models and this convinced the collaborative that it could no longer attempt to change community participation practices with the city through the city's own terms. The group shifted focus. The coalition drafted a CBA, a legally binding and privately enforceable agreement between the city, the developer, and the community. The coalition believed a CBA would be the key link to the group's power to negotiate demands. Some coalition members knew that a CBA was enacted for downtown Los Angeles during the development of the Staples Center and believed that if it happened in nearby Los Angeles, it could happen in Santa Ana. SACReD lobbied the city, urging that it pass the CBA within an intense six-month period prior to the city's voting on the SD.

Throughout the six months, SACReD evolved with hope that it could obtain a CBA. Three out of four Santa Ana city councilmembers publicly expressed support for SACReD's draft CBA, one of which made a verbal commitment. On the evening of May 20, 2010, members of SACReD headed to Santa Ana City Hall to negotiate twenty-eight demands for a proposed CBA for the Station District project.

Over sixty people attended, half of them SACReD members and allies. City officials included two city councilmembers and Community Development Agency senior staff, among other city staff. The SD developer and attorneys for both the city and the collaborative were also present. Most in attendance looked quietly during negotiations (Figure 4.1).

About midway through the negotiations, councilmember Sal Tinajero abruptly and lamentably dismissed himself. The councilmember said he had very little time prior to the negotiations to review recent revisions to the CBA draft that the collaborative had made and was therefore unprepared, the meeting had too many people, and speaking too much could potentially create negative consequences for him or the city. Yet, he also stated that the negotiations were very important because no grassroots group in the city's history had ever worked on neighborhood planning so proactively and thoroughly to negotiate, let alone attempt to come to a consensus on, a proposed legally binding agreement between the city, developer, and the community, a point he and other councilmembers had made before in other public settings. The source of discomfort in the room was his next remark: that he had qualms about some members that were part of the collaborative, that they were essentially undermining SACReD's reason for existing—responsible community development for working-poor families. While the councilmember did not name them, many present knew that he meant members of the Historic French Park Association, some of whom were present, an organization that has a contentious

Figure 4.1 SACReD negotiates for a Community Benefits Agreement with the city and Station District developer.

Source: Permission granted by Taylor & Francis to use the photograph published in González, E. R., Sarmiento, C. S., Urzua, A. S., and Luévano, S. C. "The grassroots and new urbanism: a case from a Southern California Latino community." *Journal of Urbanism: International Research on Placemaking and Urban Sustainability*, 5 (2–3), pp. 219–39.

history with the Logan barrio and others, as noted earlier. Councilmember Tinajero was aware of these battles.

Councilmember Michelle Martinez echoed similar views and also did not name the Association. She pointed out, quite perplexed and soundly, that some members' personal agendas go against SACReD's demand for additional and lower-income tiered affordable housing units in excess of what the city and developers had planned for. "They don't want affordable housing. They don't want stuff on lot 1 and that is majority for [proposed] low income housing and that should tell you that they don't have interest in your purpose," she said. In the end, these negotiations were tabled for another and smaller meeting.

Weeks after this meeting, city staff and elected officials offered an unenforceable Memorandum of Understanding (MOU) in lieu of a CBA, which the Coalition did not accept. Two councilmembers pointed out that they would not sign a CBA because SACReD included the Historic French Park Association. SACReD persisted, however, and made a final attempt to halt the Station District (SD) and persuade the city and developer to re-enter into CBA negotiations before council vote on the project. Many SACReD members and a base of supporters filled the City Council chambers when the City Council voted on the project (Figure 4.2). SACReD members and supporters wore red SACReD t-shirts, held

Figure 4.2 SACReD and allies at a City Council meeting making a final attempt to halt the Station District (SD) and persuade the city and developer to re-enter into Community Benefits Agreement (CBA) negotiations before the council votes on the SD.

Source: Permission granted by Taylor & Francis to use the photograph published in González, E. R., Sarmiento, C. S., Urzua, A. S., and Luévano, S. C. "The grassroots and new urbanism: a case from a Southern California Latino community." *Journal of Urbanism: International Research on Placemaking and Urban Sustainability*, 5 (2–3), pp. 219–39.

signs and banners advocating for a CBA, and shouted "Guaranteed benefits for the community!"

The Politics of Keeping Redevelopment Alive

The SD was approved in the summer of 2010 by the bare minimum of councilmembers when it voted four to zero, though controversy surrounded the vote (Irving 2010a). The Mayor and two councilmembers abstained in the vote, saying they had business interests in the area or received political contributions. Subsequently, the *Orange County Register* reported that two other councilpersons, both who participated in the CBA negotiations, illegally received re-election campaign donations from the Lacy housing developer prior to voting for the SD plan (ibid.). The city immediately voted to allow the councilmembers to return the funds, keeping the project alive.

But there remained suspicions in the broader community that the city had violated or was in violation of a number of redevelopment policies related to the SD project. The Santa Ana Coalition for Better Government, a group that formed around the time the SD was approved, requested that the California Controller investigate the city's Redevelopment Agency for a host of alleged violations around the SD, such as illegal use of millions to purchase and tear down homes (Irving 2011, Elmahrek 2011). These findings and allegations came at the heels of a 2010 Orange County Grand Jury finding that the city council "mismanaged behind closed doors and shielded from public review and input," when awarding a $4.85 million SD trolley contract to a company that city staff determined to be the least qualified and was a major donor to the current mayor's campaigns (Irving 2010c, Esquivel 2010). The city's prepared statement in response to the allegations said that the grand jury had found "no illegal actions" and that the city had already reinforced its contract policies (Irving 2010c).

Description of Community Organizing and Action Research

I now describe over two and a half years of community organizing and action research that culminated in the CBA demands (González et al. 2012) (see Figure 4.3). Initially, the coalition focused on coming together and representing residents in the Lacy neighborhood. As noted earlier, many organizations and stakeholders were familiar with each other, having worked together on other advocacy campaigns or cultural events in the city. General meetings and ad

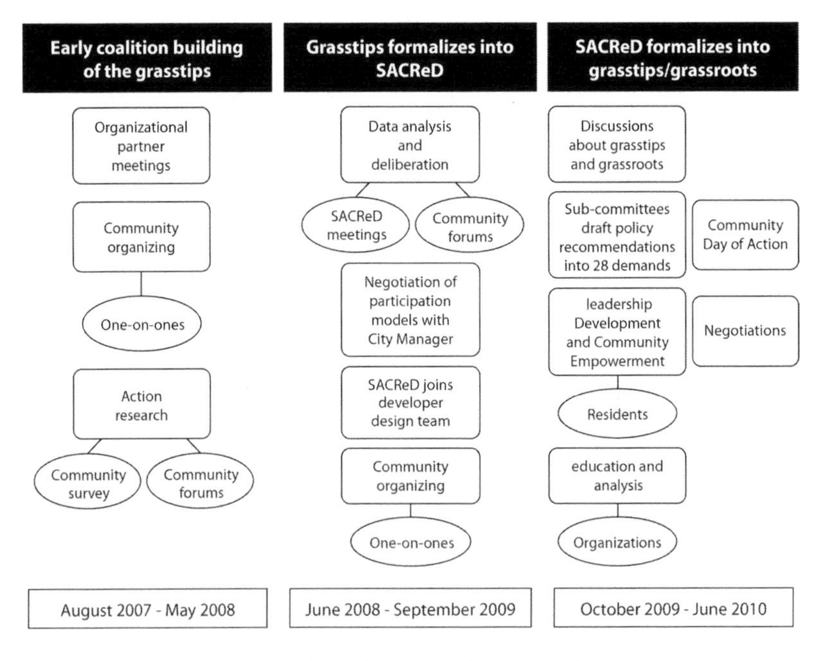

Figure 4.3 Process map of SACReD's community-based planning.

Source: Permission granted by Taylor & Francis to use the photograph published in González, E. R., Sarmiento, C. S., Urzua, A. S., and Luévano, S. C. "The grassroots and new urbanism: a case from a Southern California Latino community." *Journal of Urbanism: International Research on Placemaking and Urban Sustainability*, 5 (2–3), pp. 219–39.

hoc subcommittee meetings were the standard way the coalition came together. Members rotated responsibilities to create agendas and facilitate and translate meetings. Typical meetings spent considerable time discussing, analyzing, and interpreting the goals and objectives of the RP and SD. A lot of time was also dedicated defining the issue to analyze, detailing an agenda, exploring community-based approaches and a campaign, and conducting power analyses—plotting SD decision-makers or stakeholders who could be allies, or threats, to the group. Ad hoc subcommittees were designed to plan or carry out specific charges, such as one-on-ones with residents and councilmembers, community forums, neighborhood surveying, leadership development for residents, and analysis of data from these charges, among other activities.

Select members carried out community organizing to learn from, educate, and recruit adult Lacy residents. Coalition members who were paid through their respective nonprofit employer and some general members formed an ad hoc "organizing committee" and took charge of organization. The committee's main intentions were recruiting residents to establish and boost SACReD's resident constituency and providing them leadership development training and opportunities to practice leadership skills. The committee's additional objective was informing residents about SACReD's mission, goal, and meeting times and teaching residents about gentrification, displacement, and the SD and the RP. The organizing committee's main activities were door-to-door knocking, one-on-ones, hosting small neighborhood informational gatherings (*sesiones informativas*), large community forums, and resident meetings. The accumulation of these organizing efforts led to the creation of the "resident group," which consisted of adult Lacy residents, mostly Mexican immigrants and Spanish speakers. The group had a steady membership of about fifteen residents and they provided ideas to the organizing committee about strategies for obtaining the CBA; some of these residents participated in negotiations.

The community forums in the Lacy barrio were designed to learn from the Lacy community (see Figure 4.4). The initial forums were designed and facilitated by coalition members, including me, that did not live in the Lacy neighborhoods and subsequent forums broadened to include residents. The typical forum was designed to educate, increase critical analysis skills, obtain views from Lacy residents, and recruit and demonstrate to residents and the city the coalition's base of support. For example, most forums included SACReD PowerPoint presentations that gave a critical overview of the SD and local redevelopment politic, and outlined SACReD's intentions. In some forums members led research activities. This included facilitating workshops where residents sketched drawings of their perceived and idealized neighborhood (cognitive mapping) and small group resident discussions to further help SACReD understand community needs, concerns, and aspirations. Most of the forums were held in the Lacy neighborhood at St. Joseph Church and conducted in the evenings in Spanish and English.

SACReD deliberately searched ways to gain and maintain credibility as Lacy's "voice" and early on in the campaign this required conducting a survey. The ad hoc "research committee" designed a survey to understand the extent survey findings triangulated with all that SACReD had learned about or believed to be neighborhood

Figure 4.4 More than one hundred residents, allies, SACReD members, and others begin to fill up the parish gymnasium for a community forum days before the City Council was set to vote on the Station District. Several city officials attended the forum and expressed that they would not support a CBA. Many who were in attendance disagreed with the city officials' lack of support and responded loudly in unison with the chant "CBA all the way, CBA all the way!"

concerns and needs. The survey was administered to 48 randomly selected dwellings to understand adult resident attitudes towards and experiences with local new urbanism and transit-oriented development. I co-trained committee members and students in one of my undergraduate courses as survey administrators for survey data collection. Typically, teams of two partnered to collect the surveys.

Results of the community survey were combined with the information that all the sub-committees had collected and these results were summarized and categorized into five priority areas for the CBA: housing; cultural and historic preservation; open space; small businesses and workers; and safety. Members self-selected into five sub-committees, each named after a respective priority area, and through months of deliberation, provided specific policy recommendations.

SACReD held a carnival style half-day "Community Day of Action" event a few months before the final City Council vote on the CBA. The purpose of the Day of Action was to educate and recruit more Lacy residents, and to demonstrate to the city and neighborhood that the grassroots was well prepared to engage in urban politics. The event was held at the local elementary school parking lot in the heart of Lacy. Most booths included creative games to educate local residents and the general community about the SD plan, SACReD, and the anticipated

CBA. Individual SACReD organizations and other nonprofit groups also hosted booths to promote their work and show that they were allies with the coalition and supported the CBA.

Another key organizing process was leadership development and community empowerment. The "organizing committee" trained and educated residents and training topics included creating agendas, scheduling and facilitating meetings and forums, and mobilizing strategies. The residents were also educated on the objectives of the SD, the purpose of the CBA, definitions and availability of affordable housing in the city, and grassroots media and communications. These trained residents subsequently took on major roles, such as co-negotiating the CBA, helping schedule and moderate select meetings with councilmembers, facilitating and chairing community forums, and mobilizing friends and neighbors to action.

SACReD held education and analysis training to increase the group's knowledge around urban planning, urban policy, and community organizing. The training topics were smart growth and development, public policy, community benefits agreements, coalition development, and power analysis. The trainings required that coalition members collectively assess how much support they had from key city government officials, developers, and other sectors as well as how much political and economic influence each had on the SD.

Participation and Development Commitments

While SACReD, the city, and developers did not agree to a CBA, SACReD did achieve some noteworthy outcomes. SACReD's entire neighborhood planning, whether stemming from community organizing, action research, or debate and consensus among Collaborative stakeholders, informed the drafting of the draft CBA. For example, the CBA that the group proposed to the city depicts a deeper picture of urban development issues than the developers and city were planning for. The developer's and city's first publication of their SD concept plan draft, which was partly based on community input, conveyed key topics for redevelopment. These were park/open space for children, affordable housing and priority to Santa Ana residents, child care, local school conditions and capacity, safety and livability, retail amenities, cultural history, and family services and programs (Santa Collabortive for Responsible Development 2010). The final concept plan that was presented to the community added specificity to some of the draft topics, such as mix of for sale and for rent housing at various income levels, a community center in conjunction with school renovation, and a 1.5-acre park/open space use.

At the same time, it is difficult to separate the key issues "raised by SACReD/community" described in the concept plan and city minutes from SACReD's influence. SACReD members believed, with a high degree of confidence that the concept plan and final SD plan were highly driven by SACReD, and wondered what the final SD plan would have looked like if it had strictly resulted from the developer's and city's "community input" process, or a standard private–public planning process.

With these caveats, I describe how the approved SD plan's "commitments to Lacy neighborhood" stem from SACReD's proposed CBA. On a broader level, my emphasis is also to describe how the CBA shed light on particular issues that the approved SD does not account for. I conclude with a discussion on the extent to which the CBA matches key principles of the TOD and NU models.

A Plan for Housing

The CBA and SD plan have a similar interest in providing affordable housing, child care, relocation plans for potential displaced residents, and a residential loan program. There are, however, some fundamental differences. First, the CBA demanded that at least 50 percent of rental units be affordable to households at 80 percent Area Median Income (AMI), an "extremely low" category, and 30 percent be for households at 30 percent of AMI (Santa Collabortive for Responsible Development 2010). In terms of ownership units, the ask was 30 percent of ownership at "extremely low." The SD committed to a mix of affordable rental and for sale apartments and townhomes totaling 138 units, of which 23 are "very low" and 66 are "low."

A second difference is that the SD did not discuss how it planned to provide interim housing to avoid displacement. Instead, the city pointed out it will enforce a state mandate to provide relocation benefits. On a broader level, we can understand the CBA housing recommendations when we consider that TOD and NU projects largely neglect discussions of local political realities and housing market contexts, redevelopment for large families, and gentrification and displacement (Talen 2010).

Cultural and Historic Preservation

The CBA and the SD plan both align and differ along cultural and historic preservation dimensions. They are similar in their commitment to providing a community center, historical markers, and art in public spaces. For SACReD, the community center would provide local residents with Latina/o cultural programming and the markers and public art would highlight the contributions of Latinas/os to the development, defense, and sustainability of the area. These recommendations point to cultural and political themes common in the literature on placemaking in Latina/o communities (Rios 2009). Placemaking is also an important concept in NU, stressing open spaces and pedestrian-friendly areas to foster social interaction and walking/exercise. A fundamental difference between the CBA's and NU interpretations of placemaking is that the former reflects local conditions vis à vis cultural and political themes, while the latter reflects more general sociological and health outcomes.

Open Space

The CBA and the SD align in commitment to develop a park/open space, but the CBA detailed plans for and outcomes from use, while the SD plan did not. The CBA demanded 1.5 acres of land, and the developer promised to meet this demand. The push for a park/open space was significant. The original SD concept did not include this because establishing housing density was the priority. "At first we did not think open space was important but now realize it is," noted the developer's Project Manager, Beatrice Hsu, during a city and developer-led community meeting discussing the revised concept plan (González 2010).

The CBA envisioned particular health outcomes from the park that match common themes in NU and wished for other outcomes that do not. For instance, the CBA requires that open space include a park that would primarily service the area's children and youth population, as well as families. One specific idea for the park was increasing physical activity via soccer fields and this resonates with NU's interest in health. SACReD's push for open space, however, was also, at least, equally important for providing local residents opportunities to engage in placemaking. This includes performance space for local youth and resident-led cultural and sports events, and a community mural.

The push for open space was also about creating retail options that could appeal to the area's demographic and economic livelihood, but these were not part of the SD plan. The CBA required the developer to lease commercial space at discounted rents to qualifying small, locally owned, neighborhood-serving retail establishments, with priority given to businesses operated by SD residents. The major intention is to maintain and encourage businesses that market to the current and majority Mexican working-class population, thereby sustaining a majority Mexican cultural identity in the area, hence helping reduce patterns of displacement and gentrification. The developer responded that the SD is not a retail-driven project, making the point null and void.

Public Safety

The CBA recommended that redevelopment commit programs and policies designed to address a range of public safety concerns and the SD omits this. Some of the major ideas included creating a neighborhood watch program, stricter enforcement of liquor license laws, and built environment and general design elements to help protect pedestrians and school children, generally. Safety was a concern because of crime and potential alcohol-related delinquency that may result from new or future bars and restaurants in the downtown. Generally, NU does not include attention to these types of safety issues.

Coalition Challenges

Disagreements within SACReD surfaced and were resolved over different assumptions about "resistive praxis." Martinez (2010) defines resistive praxis as "the set of practices and beliefs that emerges from shared experiences and history in an urban neighborhood and informs the content and methods of local political practice" (p. 41). The disputes were between some seasoned Santa Ana activists and members representing some nonprofits, all who did not live in the Lacy neighborhood, hereinafter referred as the "grasstips," and another internal group comprised of a mix of a younger generation of activists and seasoned activists.

The "grasstips" believed it could suitably make all SACReD decisions and represent Lacy's interests, while other members argued that the "grassroots," local residents who were mainly predominantly Spanish-speaking Mexican immigrants living in the Lacy neighborhood, should be better integrated into the group to assume roles and responsibilities equally.

The Coalition went through rather routine deliberations, defining and practicing resistive praxis. Many of the Latina/o activists who organized the early coalition-building meetings had a common resistive praxis developed from years of struggle on other city and countywide issues, as noted earlier. These activists were ready to be accountable to the Lacy neighborhood by listening and learning from and representing the area. We see this commitment during the coalition-building stages when some SACReD members began meeting with some Lacy neighborhood residents.

While SACReD had transcended much of the criticisms that it had directed toward formal public–private participation that helped give way to the SD, the Coalition was also perpetuating some of the shortcomings that it identified. For example, the community residents that SACReD was so passionate about lived in the Lacy neighborhood. The Lacy neighborhood largely consists of immigrant and first-generation Mexican-American residents. Many in SACReD, while having grown up in other Santa Ana barrios of previous generations and being first-generation born of Mexican immigrant parents, had "shared experiences and history" with Lacy that were symbolic at best. During SACReD's first two years of existence, only one Lacy neighborhood resident participated in the group's meetings, and his involvement was scarce and sporadic, with no involvement in decision-making. Some members did not believe the low level of representation and decision-making from Lacy was an issue. These members questioned the need and practicality of involving residents in day-to-day decisions. They reasoned that the existing SACReD membership was well informed about Lacy's needs, having already conducted community organizing and action research in two years, and capable of making all organizational decisions. Other members, especially those from the organizing committee, cordially disagreed with these assumptions. Through deliberation and consensus, the entire collaborative agreed to broaden the group's dominant "resistive praxis." The organizing committee developed an ad hoc "resident group" to reflect this stance. Residents attended a few coalition meetings to share their ideas about strategy for obtaining the CBA. For much of the time, organizers advocated within the Coalition that these strategies be included and facilitated from time to time general meetings to ensure residents had significant decision-making power within the group. Subsequently, residents co-designed and facilitated negotiations with the city.

Neighborhood conflicts and sub-group-interests tested SACReD's unity and goals. A group of White middle-class homeowners, some of which were residential investors from French Park, started attending meetings about two years after SACReD formed. The group was comprised of home preservationists with a vested interest is retaining historical meaningful existing single-family dwellings in the SD. As noted earlier, there were some personal lingering ill feelings among some SACReD members who represented Logan toward these residents. Later, the CBA represented compromises regarding the balance of preserved homes and new affordable housing in the SD.

There were also strong disagreements between the coalition and the developer/city over housing and open-space land use. Much of the land in the SD was purchased with redevelopment dollars, triggering a requirement for one-to-one replacement of the housing. Yet, the city pointed out that the CBA's demand for 1.5 acres of park space would subtract land that was intended by the city to meet this replacement mandate. The most influential housing advocate on the Coalition expressed support for this CBA demand, but his final reticence at the drafting of the CBA was based on the decreased number of affordable housing units that these demands would require. The affordable housing advocate stood steadfast by the SD one-to-one replacement and dissuaded the Collaborative from suing the city when the project was approved. In the end, the Collaborative opted to remain united and accepted the city's promises.

References

City of Santa Ana. 2009. Request for Qualifications: The Station District [online]. City of Santa Ana.

City of Santa Ana. 2010. Public Hearing for Transit Zoning Code (no. 84) and proposed development in the Station District.

City of Santa Ana. 2012. Press Release. City of Santa Ana Redevelopment Agency and Related California. City of Santa Ana Redevelopment Agency and Related California break ground on Station District apartment homes.

Diaz, David R. 2005. *Barrio urbanism: Chicanos, planning, and American cities*. New York: Routledge.

Elmahrek, Adam. 2011. "Latino groups call for grand jury investigation of city officials." *Voice of OC*, January 6, 2011.

Esquivel, Paloma. 2010. "O.C. grand jury criticizes Santa Ana council over transportation contract." *Los Angeles Times*. May 25, 2010. http://articles.latimes.com/2010/may/25/local/la-me-oc-grand-jury-20100525.

González, Erualdo. 2010. Meeting notes from the City of Santa Ana's community meeting on the Station District initial concept plan.

González, Erualdo R., Carolina Sarmiento, Ana Urza, and Susan Luévano. 2012. "The grassroots and new urbanism: A case from a Southern California Latino community." *Journal of Urbanism: International Research on Placemaking and Urban Sustainability* 5 (2–3):219–39.

Greenhut, Steven. 2008. "Is it all immigrants' fault?" *The Orange County Register*. March 2, 2008. www.ocregister.com/opinion/immigration-illegal-immigrants-1990890-born-people.

Irving, Doug. 2010a. "Campaign money raises questions over Santa Ana vote." *Orange County Register*. September 14, 2010. http://m.ocregister.com/news/vote-266466-city-council.html.

Irving, Doug. 2010b. "High rises and more homes in Santa Ana." *Orange County Register*, January 11, 2010. www.ocregister.com/articles/new-228495-city-downtown.html.

Irving, Doug. 2010c. "Well-connected firm got flawed contract." *Orange County Register*. May 24, 2010. http://articles.ocregister.com/2010-05-24/news/24636798_1_grand-jury-cordoba-vincent-sarmiento/2.

Irving, Doug. 2011. "Groups seek grand jury probe of Santa Ana." *Orange County Register*. January 5, 2011. www.ocregister.com/news/coalition-282858-grand-jury.html.

Martinez, Miranda J. 2010. *Power at the roots: community gardens, and the Puerto Ricans of the Lower East Side*. Lanham, MD: Lexington Books.

Related, Griffin Realty Corporation and WHArchitects. 2010. The Station District.
Rios, Michael. 2009. "Public space praxis: cultural capacity and political efficacy in Latina/o placemaking." *Berkeley Planning Journal* 22 (1):92–112.
Santa Ana Collaborative for Responsible Development. 2010. Proposed business points for Community Benefits Agreement (or Agreements) between SACReD, City of Santa Ana Related Companies relating to Phase 1 of Station District and development occurring in Transit Zoning Code Area. Santa Ana, CA.
Talen, Emily. 2010. "Affordability in new urbanist development: principle, practice, and strategy." *Journal of Urban Affairs* 32 (4):489–510.
United States District Court, Central District of California, Southern Division. 2004. Josephine Andrade et. al., Plantiffs, v. Miguel A. Pulido, Mayor of City of Santa Ana, in his official capacity, et. al., Defendants. edited by Clerk, U.S. District Court Central District of California Santa Ana Office.
Villa, Raúl H. 2000. *Barrio-logos: space and place in urban Chicano literature and culture*, 1st ed. History, culture, and society series. Austin, TX: University of Texas Press.

Conclusion

This book sheds light on the linkages between spatial alienation in downtown central city redevelopment and the grassroots in defense of space. My intention has been to interrogate some of the normative assumptions that have supported the common practice redevelopment, assumptions the mainstream urban planning literature often leaves unquestioned and even unspoken. Redevelopment in Santa Ana has ignored the fact that the downtown has been both culturally affirming and affordable for Mexican working-class immigrants; a vibrant and functioning community has been socially constructed as blight. My examination of this social construction and the implementation of practices based on it has required exposing the hidden mechanisms of redevelopment. I have also sought to tell the story of how Santa Ana's grassroots responded to the threat of spatial alienation as well as its implementation. By revealing how the grassroots went about unfettering instead of accepting redevelopment plans and projects, we learned the achievements that these grassroots endeavors produced and the challenges they encountered.

In Santa Ana, racism and elitism are very connected and have justified and promoted key redevelopment plans and projects for more than forty years, although the rhetoric has changed over time. By the mid-1970s, I found that city officials had recognized that Santa Ana was rapidly growing in numbers, browning, and that working-class and poor people were becoming the face of the city. The Mexican population was growing to the point where in 1980 the city was no longer majority White—but the city officials were virtually all White. As the City Council planned for the city's future, they hoped for a return to a future more like Santa Ana's early twentieth-century White past. Thus, they sought to craft spaces that appealed to their own social group, assuming they could systematically cast away the resentment of Santa Ana's actual residents. Racist and elitist ideologies appeared both overtly and covertly in authoritative redevelopment documents and community-wide publications that cite and promote them as early as the 1974 DSADP. We learned that the racist dimension to this discourse typically reduced nuanced cultural, social, and economic realities of downtown everyday and commercial life into a few sensationalist and hegemonic characterizations. The City Council and redevelopment officials saw the Mexican immigrant consumer, resident, and business sector as occupying space that, if they were to relinquish it, would return the town to the privileged position it had in the region decades earlier. Investors and other stakeholders were ready to reclaim this space after a period of private and city disinvestment. The narratives they employed described a marginal and abominable group of Mexican consumers and residents that were flooding the downtown area and an abundance of low-quality businesses that served them. As a whole, such elitism and racism serve as undercurrents to planners' fears and resentments about different social and cultural groups (Harwood 2012).

Latino City vs. the Santa Ana City Council

City plans did not translate easily into practice. Santa Ana grew fast through the 1970s, with Latina/o fast dominating the population. Fourth Street became *La Cuatro*. Once patterned with vacancies, *La Cuatro* bloomed with new businesses, many locally owned, appealing to Mexican immigrants who were the area's sustaining consumer. These changes were at odds with the plan the City Council had created for the city; even the use of and presumed benefits of eminent domain to accomplish changes did not materialize completely.

By the mid-1980s, public officials began to recognize that a return to the Santa Ana of the 1950s was unlikely. The racist and elitist narrative rescinded a bit, and narratives of embracing cultural, economic, and democratic inclusion began to emerge. *La Cuatro*'s growing economic success supported the creation of a public–private partnership to develop Fiesta Marketplace on the eastern edge of *La Cuatro*. This culturally themed redevelopment offered affordable amenities and consumption options with a Mexican and working-class consumer in mind. It represented a signal of commitment to downtown commerce interventions that reflected the city's growing Mexican families, immigrant base, and low-wage constituency.

If Santa Ana's Mexican activists were hopeful that Fiesta Marketplace signaled an abandonment of racist and elitist rhetoric, the 1990s brought disappointment. Now, the narrative included redevelopment language that borrowed from creative class (CC), new urbanism (NU), and transit-oriented models (TOD), repackaging traditional racist and elitist views. Focused on the arts, creative types, and work-lift housing redevelopment, the city sought yet again to whiten its downtown.

At this point there was no question of abandoning *La Cuatro* altogether. *La Cuatro* was booming. For the time being, at least, public–private interventions could not eliminate it. The city therefore created the Artists Village, a commercial and housing space area on one contiguous edge of *La Cuatro*. The objective of the new urbanist work-live loft redevelopment was to attract and sustain a mix of creative and upwardly mobile consumers and residents from the region. If they would not attempt to dismantle *La Cuatro*, they also would make no investment in it or the poor, working-class, and immigrant Latina/o that patronized it.

Newer en vogue redevelopment models sweeping the nation helped shift the city's narrative about what constitutes a desired community. The rhetoric shifted more to emphasizing class markers than ethnic or racial ones. The idealized community that new urbanism (NU), creative city (CC), and transit-oriented redevelopment (TOD) approaches envisioned aimed to attract sub-demographic groups with privilege, often labeled "diverse" and "hip": people who enjoy shopping and consuming in mid- and higher-end establishments who have expendable income to eat at trendy local restaurants, and interest in perusing galleries and patronizing coffee shops, bars, and cocktail lounges. Many other central city areas were prioritizing this demographic as well. By outlining these trends, we understand that the new redevelopment plan targeted, rather than only the White and middle- and upper-class gentry they had sought in the previous thirty years, a broader demographic group (Lejano and González 2016). This constituted a fluid intersection of race, class, citizenship, culture, and related categories to bring to the forefront the concept of privilege.

The Accomplishments and Limitations of the All-Latina/o City Council

This book has addressed the role of race on the City Council in shaping spatial alienation in Santa Ana. Historically, White men dominated the City Council. Latinas/os were heavily underrepresented for decades even as Latinas/os increasingly dominated the city itself. In the 1970s and early 1980s, Latina/o candidates represented not only racial and ethnic diversity on the City Council, but a point of view for some activists that government should support the well-being of working-class and poor communities.

Over time, however, the sense that a councilmember's racial and ethnic identity predicted he or she would fight for the socioeconomic interests of the poor, working-class, Mexican immigrant, Mexican-American, and Chicana/o residents of Santa Ana began diminishing among activists. By the time the city achieved an all-Latina/o city council in the mid-2000s, activists no longer had expectations that such a council would fight for their interest. The highly contested downtown redevelopment projects that ensued bore out their low expectations.

The present day redevelopment models and racial-ethnic urban governance context in Santa Ana add new perspectives to the Latino urbanism literature. This body of literature generally highlights historical examples illustrating that elite Anglo policy actors perpetuate discriminatory redevelopment practices in marginalized communities of color (Diaz and Torres 2012, Irazábal and Farhat 2008). The 1970s and early 1980s redevelopment era in the central city of Santa Ana was in many respects not all that different from other heavily Latina/o-populated central cities at that time. Diaz (2005) observes that redistribution policy has historically favored civic center zones and subsidized stadiums, hotels, and enclave development for professional classes over Mexican immigrant and Chicana/o barrios struggling to maintain a just baseline of affordable housing, infrastructure maintenance, and inclusionary planning. As this book has shown, Santa Ana activists and residents historically fought for Latina/o representation in the City Council in the hopes that such leadership would empathize and channel policy and resources designed to ameliorate concerns about the Mexican immigrant and Chicana/o barrios and *La Cuatro* downtown commercial district. In many instances, the struggle extended to fighting against eminent domain, a major threat to downtown barrios and *La Cuatro*. From the 1990s to the mid-2000s, urban governance began to shift. When a Latina/o member filled the last council seat, Santa Ana became the first major U.S. city to have an all-Latina/o city council. Over the decades, the hope among seasoned and younger activists has been that racial and ethnic political representation would advocate, vote for, and wield resources to directly and primarily benefit the city's poor, working-class, and immigrant residents whose lives are largely bound in and around the downtown area. Regardless of decade and revitalization strategy, redistribution of policy and resources for downtown commercial and housing options continued to favor the middle-class and native-born Whites, whether it terms them the creative class, "artsy," or "hip."

Diaz (2012) writes that new urbanism has stripped away affordable housing in many cities, causing gentrification and exclusion of the working class and the poor (p. 38). This book bears out this statement through its policy analysis of the new urbanist Renaissance Specific Plan (RP), exposing the alienating nature of new urbanism as communicated in the plan. The plan as a whole helped the city to justify and impose alienating cultural and class redevelopment onto existing urban spaces and many of those ideas are highly visible in downtown redevelopment. The Foucauldian analysis of the RP reveals that disciplinary mechanisms are embedded extensively, from zoning to permitting, and that they manifest what Foucault termed "governmentality" (Foucault 1991). This means that within the RP we find a set of practices that reflect the art of government planning which set the stage to spatially alienate the downtown and barrios of Santa Ana.

Community Participation and SACReD

This book's case study also provided insights on community participation. The 1974 DSADP was a classic example of a plan created with rational planning principles. The plan was prepared through interviews with relevant city departments and developers, field observations of the downtown area, and a review of existing data sources. The resident constituency and downtown merchants had no input in the process. The obvious implication of this approach is that the needs and desires of those who were increasingly making up a significant block of the living and consuming were deliberately excluded in an important plan. In this respect, Santa Ana planning was another example of rational planning practices that dominated during the 1960s and 1970s. In those eras, professional and paid experts relied on traditional planning techniques and research to plan, and the broader community was often relegated to tokenism participation, or no involvement. In 1969, for example, in seminal research, Arnstein (1969) showed that power sharing among residents was very low when state actors led or sponsored participation activities in planning for urban renewal and related social programs in central cities. Arnstein concluded that low levels of power sharing dominated many cities across the country and that higher levels of power sharing typically follow only when the citizens take action—not because the city grants them a voice without conflict.

My analysis suggests that the Santa Ana 1970s and early 1980s redevelopment era conforms to Arnstein's general observations, especially in terms of Latina/o community participation. In Santa Ana, the grassroots proactively questioned the city's lack of power sharing over redevelopment decisions. But it was not until the community took action to take redevelopment power from city officials that power relationships altered. Protest in public spaces, testimony in city hall, and participation in city-sponsored forums and meetings produced some level of change.

Nationwide, a body of research following Arnstein has shown that neighborhood organizations have been using different types of community-based approaches to create power-sharing relationships with city governments. Ranges of examples have shown that groups plan along or against local government in pursuit of redistributive and participatory policies and practices and that some are successful (see for example Clavel 2010, Smock 2004, Gittell and Vidal 1998). Arnstein's critiques of participation practices continue to have relevance in private–public participation practices that are commonly used to prepare contemporary new urbanist and transit-oriented development plans. The SACReD experience revealed that a city and developer partnership with proactive, intense, and compressed community participation activities might not signal the city's willingness to share decision-making power with residents and activists. This finding resembles that of many other studies that find participatory planning methods do not always coincide with a willingness to let residents influence policy outcomes (see Bond and Thompson-Fawcett 2007, Day 2003, Toker and Pontikis 2011). SACReD recognized that it would need to alter the ways the city was governed if we were to gain the voice we sought. SACReD's grassroots approach was about creating new forms of governance to maximize the chances that there would be power sharing with which the group could be satisfied. The collaborative in some respects did alter how the city governed the community development planning process. SACReD introduced and took command of a new set of community participation activities and introduced negotiations for a legally mandating community benefits agreement (CBA), a historic grassroots accomplishment in the city.

SACReD's approach was evolutionary. Coming to the point of negotiating the CBA took years. Part of this evolutionary nature was tied to the city's political environment. At times the city would make a request or impose a deadline that necessitated a quick response and the organization's allocation of time had to change. Collaborative members have to call unexpected meetings to respond to these new requests and deadlines; sub-committees would form to take on particular roles and responsibilities. Collectively, my observations of SACReD's work contrasts with much of the NU and general urban planning literature, which describes innovative methodologies to engage communities in time-compressed and apolitical planning processes. In accordance with Arnstein's findings, SACReD had to take power instead of hoping the city would relinquish it through such processes.

Directions for Future Research

This book included one in-depth case study and the extent to which some findings may apply to multiple contexts may vary. An intention of the book was to provide what Geertz (1973) calls a "thick description" and suitably allow readers to make associations between elements and findings of this research and their own situations. I suspect some of my findings are transferable to other cases or that they provide useful insight for understanding and examining similar urban contexts.

Future research examining cores of central cities with a high Latina/o population may benefit from questions that consider redevelopment schemes that dominate present-day urban planning. Studies could examine the extent to which public–private partnerships carrying out any of these redevelopment approaches are receptive to broadening community participation beyond standard processes, and what difference this may make in terms of policy outcomes and implementation. To what extent do public–private partnerships incorporate community-designed and community-based participatory methodologies, and what policy recommendations and policy implementation can be traced to any of these practices?

The SACReD case also provides insights related to redevelopment CBA negotiations. SACReD was unable to obtain a legally binding redevelopment agreement with the city and developer. SACReD contends that the main reason for this was a strategic error we made of negotiating with the city instead of the developer. The group's retrospective insight is in line with Baxamusa's (2008) research suggesting that community groups tend to be more successful in obtaining a CBA when they negotiate directly with the developer instead of negotiating with city staff or politicians. Part of the reason the group focused on the city was the consistent overlap between the roles and responsibilities of the city and the developer in conducting outreach and finalizing the plan. These overlapping roles made it difficult for the coalition to distinguish who was setting the objectives for and carrying out the Station District (SD) planning community meetings. As a whole, the group did not entertain the idea that the developer would wield the greatest political clout.

The SACReD case also provides lessons that could be useful for future research concerned with community participation in lower-income and Latina/o neighborhoods. The varied cultures and growing income inequities that characterize U.S. society almost guarantee that some planners and developers preparing redevelopment projects will work outside of familiar cultural and economic contexts at some point in their careers. For SACReD, cultural competence was critical to gathering and interpreting data from the local neighborhood. This skill also contributed significantly to the depth and breadth of community engagement that NU public participation exercises have historically lacked (Day 2003). Culturally competent organizers made it possible for Santa Ana residents to have a voice in the development of their own neighborhoods.

The Future of Urban Planning in a Latinizing United States

Present redevelopment trends of the central city urban center have a vision to "revitalize" and "diversify" spaces. We see reclaiming in practice in the many cities whose governments seek to encourage populations to return to under-invested downtown cores. New urbanism (NU), creative class (CC), and transit-oriented principles characterize much of these efforts. The Charter of New Urbanism's catchphrase to "reclaim" cities and the creative class movement toward reimagined communities of privilege become especially chilling to working-class Latinas/os and working-class communities of color in the central city. The use of this type of messaging and redevelopment models to pursue such gentrification goals for downtown cores is often at odds with immigrant, ethnic, and working-class and working-poor populations who continue to live and consume in the central city. They are also among the most vulnerable to displacement by reclamation by the urban elites city officials seek to bring back to urban cores. Urban governance and political and economic interests are intertwined with immigrants and the working class claiming rights themselves. We can turn to *Latino City* to help us understand that the future of urban planning for downtowns in the central city throughout the United States is already here.

References

Arnstein, Sherry. 1969. "A ladder of citizen participation." *Journal of the American Institute of Planners* 34 (4):216–24.

Baxamusa, Murtaza H. 2008. "Empowering communities through deliberation: the model of community benefits agreements." *Journal of Planning Education and Research* 27 (3):261–76.

Bond, Sophie, and Michelle Thompson-Fawcett. 2007. "Public Participation and new urbanism: a conflicting agenda?" *Planning Theory & Practice* 8 (4):449. doi: 10.1080/14649350701664689.

Clavel, Pierre. 2010. *Activists in city hall: the progressive response to the Reagan Era in Boston and Chicago*. Ithaca, NY: Cornell University Press.

Day, Kristen. 2003. "New Urbanism and the challenges of designing for diversity." *Journal of Planning Education and Research* 23 (1):83–95. doi: 10.1177/0739456X03255424.

Diaz, David R. 2005. *Barrio urbanism: Chicanos, planning, and American cities*. New York: Routledge.

Diaz, David R. 2012. "Barrios and planning ideology: The failure of suburbia and the dialects of new urbanism." In Diaz, David R. and Torres, Rodolfo D. (eds.) *Latino urbanism: the politics of planning, policy, and redevelopment*. New York: NYU Press.

Diaz, David R., and Torres, Rodolfo D. 2012. "Introduction." In *Latino urbanism: the politics of planning, policy and redevelopment*, 1–20. New York: NYU Press.

Foucault, Michel. 1991. In Burchell, Graham, Gordon Colin, and Miller Peter (eds.) *The Foucault effect: Studies in governmentality*. Chicago: University of Chicago Press.

Geertz, Clifford. 1973. *The interpretation of cultures: selected essays*. New York: Basic Books.

Gittell, Ross J., and Avis Vidal. 1998. *Community organizing: building social capital as a development strategy*. Thousand Oaks, CA: Sage Publications.

Harwood, Stacy A. 2012. "Planning in the face of anti-immigrant sentiment: Latino immigrants and land-use conflicts in Orange County California." In Rios, Michael, Vazquez, Leonardo, and Miranda, Lucrecia (eds.) *Diálogos: Placemaking in Latino Communities*. New York: Routledge, pp.36–49.

Irazábal, Clara, and Ramzi Farhat. 2008. "Latino communities in the United States: placemaking in the pre-World War II, postwar, and contemporary city." *Journal of Planning Literature, February 2008*, 22 (3):207–28. doi: 10.1177/0885412207310146.

Lejano, Raul P., and González, Erualdo R. "Sorting through differences: the problem of planning as reimagination". *Journal of Planning Education and Research* 0739456X16634167, first published on March 23, 2016 as doi:10.1177/0739456X16634167.

Smock, Kristina. 2004. *Democracy in action: community organizing and urban change*. New York: Columbia University Press.

Toker, Zeynep, and Kyriakos Pontikis. 2011. "An inclusive and generative design process for sustainable urbanism: the case of Pacoima." *Journal of Urbanism* 4 (1):57. doi: 10.1080/17549175.2011.559956.

Index